T0297195

Springer Water

The book series Springer Water comprises a broad portfolio of multi- and interdisciplinary scientific books, aiming at researchers, students, and everyone interested in water-related science. The series includes peer-reviewed monographs, edited volumes, textbooks, and conference proceedings. Its volumes combine all kinds of water-related research areas, such as: the movement, distribution and quality of freshwater; water resources; the quality and pollution of water and its influence on health; the water industry including drinking water, wastewater, and desalination services and technologies; water history; as well as water management and the governmental, political, developmental, and ethical aspects of water.

More information about this series at http://www.springer.com/series/13419

Anna Entholzner · Charles Reeve
Editors

Building Climate Resilience through Virtual Water and Nexus Thinking in the Southern African Development Community

 Springer

Editors
Anna Entholzner
Climate Resilient Infrastructure
  Development Facility
Pretoria
South Africa

Charles Reeve
Climate Resilient Infrastructure
  Development Facility
Pretoria
South Africa

ISSN 2364-6934   ISSN 2364-8198   (electronic)
Springer Water
ISBN 978-3-319-28462-0   ISBN 978-3-319-28464-4   (eBook)
DOI 10.1007/978-3-319-28464-4

Library of Congress Control Number: 2016936962

Disclaimer: The British Government's Department for International Development (DFID) financed this work as part of the United Kingdom's aid programme. However, the views and recommendations contained in this communication are those of the individual authors, and neither DFID nor CRIDF is responsible for, or bound by the recommendations made. Similarly, the views expressed are those of the individual authors and do not necessarily reflect DFID or CRIDFs views.

Printed on acid-free paper

This Springer imprint is published by Springer Nature
The registered company is Springer International Publishing AG Switzerland

In memory of Dave Phillips and Lawrence Musaba.

Thought leaders, respected colleagues, and much loved friends.

Foreword

In January 2015, the doomsayers were vindicated when the World Economic Forum (WEF) released its 10th Global Risks Report. The report, which aims to shed light on global risks and help create a shared understanding of the most pressing issues, the ways they interconnect and their potential negative impacts, draws on the inputs of over 900 public-sector as well as private-sector stakeholders globally. The 28 global risks are ranked and mapped according to two dimensions—likelihood and impact. Over the past decade the risks deemed to pose the greatest negative impact on the global economy have included inter-state conflict, financial crises, energy prices; and the spread of infectious diseases. This year's report, for the first time, places water as the number one risk.[1]

Now that the wise and wealthy of the WEF have taken water risk seriously there is an increased recognition of the crucial role water plays lubricating the wheels of the global economy and social systems across all levels of scale. The realisation that water issues pose an existential risk to their operations has caused captains of industry to initiate programmes on water security (going beyond improving water-use efficiency within their own operations and extending to catchment conservation efforts in some parts of the world); and in water-stressed parts of the world (predicted to be home to half the world's population by 2030) water is an issue ascending the political agenda. In the SADC region water issues have made it into the regional policy structures, with talk of desertification, the el Nino effect, climate change, droughts and floods all having a regional resonance. But what about the ordinary people living in the region, the men, women and children, for whom water scarcity is an all too daily reality? How do they fit into this picture of water as a global risk factor?

It turns out that where they fit into the broader picture is largely an accident of birth—where they live will to a large degree determine how much water they can

---

[1]In the 11th WEF Global Risks Report released in January 2016 water issues have slipped to number three in the list of 'Global risks in terms of impact'. The number 1 risk in the 2016 report is 'Failure of climate change mitigation and adaptation'. Water issues have been in the top three of global risks for the last 5 years.

access at various levels to supply their direct as well as indirect needs. The north-west parts of southern Africa enjoy high levels of rainfall, delivered in a relatively steady pattern. Rainfall becomes more variable as one moves east and drops dramatically as one moves south. What this book does is to analyse the impacts of this regional distribution of rainfall in relation to economic development patterns, urbanisation and water consumption for agriculture and energy production. Beyond the basic daily amount of water each human needs for their survival (depending on your habits this is deemed between 25 and 50 litres a day) the vast consumption of water is through the provision of products and services. When we speak of the region facing a future of water scarcity it is the ability to grow crops, develop industry, generate electricity and sustain ecosystems where the challenge lies for water managers. If the SADC region is going to become more economically developed, less unequal and ultimately more sustainable, it will need well-managed water supplies as an input into a range of products and services, or else water will truly become our number one regional risk.

I first heard the term "virtual water" as a postgraduate student majoring in Earth Observation and Remote Sensing at the University of London. My immediate interpretation was that this was in some way "digital water"—a new way of classifying water resources on the flickering screen of my computer where a GIS programme could plot its supply, flux and use on the ground. It turns out that my understanding of the term "virtual water" was wrong, but my understanding of the concept and what it could teach us about regional and global flows, shifts, opportunities and risks was in the right spirit. If water is a fundamental factor in the production of most of the products we consume and trade today, then better understanding of the production and, more importantly, the trade in these products is a vital part of our regional water picture. Over time I came to realise that our economies and ultimately our people in this region are inexorably bound by these flows, most of which take place spontaneously driven by private interest and facilitated by the market. Arguably these flows have done more for regional integration and development than any number of master-plans or infrastructure projects.

This book makes a much-needed contribution to analysing these flows and mapping their impacts and opportunities for the region, a risk-mitigation strategy if ever there was one. It is time for Virtual Water to become real as evidenced by the statement made by the President of the African Development Bank, Donald Kaberuka, describing the Kazungula Bridge over the Zambezi (linking Zambia and Botswana) as the single most important project which the bank is financing on the continent today. This bridge will reduce the transit time between the two countries from the current 36 hours to around an hour; and serve to link the relatively water-rich agricultural areas of western Zambia to the water-scare regions and markets to the south. Perhaps the President knew that for every tonne of maize being carried by the trucks which one-day will cross that bridge almost a thousand cubic metres of water will change hands (virtually of course)?

Anton Earle
Stockholm International Water Institute

Preface

The Climate Resilient Infrastructure Development Facility

At the request of the Southern African Development Community (SADC) the Climate Resilient Infrastructure Development Facility (CRIDF) programme was introduced by the United Kingdom to support the implementation of water infrastructure in the SADC region as outlined in the Regional Infrastructure Development Master Plan (RIDMP) and related SADC documents.

CRIDF promotes the delivery of small- to medium-scale infrastructure across SADC through technical assistance aimed at developing sustainable pro-poor infrastructure projects, and facilitating access to finance to deliver the infrastructure. The Facility, through this infrastructure, aims to build climate resilience for poor communities across the SADC region, to enhance cooperation in shared river basins, and to build an evidence base for the national and regional benefits of cooperation.

However, CRIDF recognises that small- to medium-scale infrastructure has limited impact on transboundary water management and the building of climate resilience. It was with this in mind that the most strategic theme was introduced to clearly articulated arguments for changing the way infrastructure is planned, designed, delivered and operated to build climate resilience that will address poverty and build regional cooperation.

This strategic theme seeks to address the above challenges with the following mechanisms:

- Improved information/evidence will lead to "better" decisions even in the absence of effective and inclusive decision-making processes;
- Developing networks around climate resilience will stimulate engagement by interest groups in society (climate change NGOs, citizen groups, local governments and the media which enable them to advocate for pro-poor, climate resilience) within their own national policies;

- Brokering relationships between local as well as regional decision-makers, local governments' community representatives, NGOs and the media. This should lead to transformational change with regard to these relationships and how knowledge/information about the planning, implementation and operation of water infrastructure is disseminated;
- Empowering stakeholders through a better understanding of climate resilience issues which will result in improved decision-making both for local needs and regional priorities;
- Creating momentum in the public and in the civil society while increasing transparency will contribute to transformational change within SADC.

It is within the context of this strategic theme that the contributions covered in the book were prepared; these aim at assessing the extent to which Virtual Water, water footprint, and the water–food–energy nexus concepts could contribute to SADCs overarching integration, poverty reduction, peace dividends and economic growth objectives.

Anna Entholzner

Contents

Contributors

John Anthony Allan Department of Geography, King's College, London, UK

Hannah Baleta Pegasys, Cape Town, South Africa

Stephen Boyall Adam Smith International, London, UK

Anna Entholzner Climate Resilient Infrastructure Development Facility, Pretoria, South Africa

Simon Krohn Simon Krohn Consulting Pty Ltd, Hobart, Australia

Simbarashe Mangwengwende Zambezi Hydro Power Company, Harare, Zimbabwe

Mike Muller Wits University's School of Governance, Johannesburg, South Africa

Lawrence Musaba Director Southern African Power Pool Coordination Centre, Harare, Zimbabwe

Guy Pegram Pegasys, Cape Town, South Africa

David Phillips Adam Smith International, London, UK

Charles Reeve Climate Resilient Infrastructure Development Facility, Pretoria, South Africa

Barbara Schreiner Pegasys Institute, Cape Town, South Africa

Reginald M. Tekateka Technical Advisory Committee African Minister Council of Water, Johannesburg, South Africa

Anthony Turton Centre for Environmental Management, University of Free State, Bloemfontein, South Africa

Acronyms

ABCD	ADM, Bunge, Cargill, Dreyfus
AfDB	African Development Bank
ASR	Aquifer Storage and Recovery
AECF	African Enterprise Challenge Fund
AMCOW	African Ministers Council on Water
ARWR	Annual Renewable Water Resources
AWF	African Water Facility
BRICS	Brazil, Russia, India, China, South Africa
CAADP	Comprehensive Africa Agriculture Development Programme
CAB	Congo Air Boundary
CAP	Common Agricultural Policy of the European Union
CAPP	Central African Power Pool
CDP	Carbon Disclosure Project
CEO	Chief Executive Officer
CRIDF	Climate Resilient Infrastructure Development Facility
CSR	Corporate Social Responsibility
CTC	Central Transmission Corridor
DAM	Day Ahead Market
DDT	Dichlorodiphenyltrichloroethane
DfID	Department for International Development
DRC	Democratic Republic of the Congo
EAPP	Eastern African Power Pool
EU	European Union
EUWF	European Union Water Facility
ENSO	El Niño Southern Oscillation
FAO	Food and Agriculture Organisation
FLS	Front Line States
GCM	Global Climate Model
GCMs	General Circulation Models
GDP	Gross Domestic Product

GWP	Global Water Partnership
HDI	Human Development Index
HPI	Happy Planet Index
IAM	Integrated Assessment Modelling
IBT	Inter Basin Transfer of Water
IGMOU	Inter-Governmental Memorandum of Understanding
IGWAC	International Ground Water Assessment Centre
ICP	International Cooperating Partners
IFAC	International Federation of Accountants
IMF	International Monetary Fund
IPP	Independent Power Producer
IRP	Integrated Resources Plan
ISIC	International Standard Industrial Classification
ITC	Independent Transmission Company
ITCZ	Inter-Tropical Convergence Zone
IUMOU	Inter-Utility Memorandum of Understanding
IWMI	International Water Management Institute
JWC	Joint Water Commission
LCA	Life Cycle Analysis
LIMCOM	Limpopo Watercourse Commission
LUSIP	Lower Usuthu Smallholder Irrigation Project
NWRS	National Water Resource Strategy
MAP:MAR	Mean Annual Precipitation to Mean Annual Runoff Ratio
MRC	Mekong River Commission
NEPAD	New Partnership for African Development
NEXAS	Natural Ecosystems Expect Accountable Stewardship
NGO	Non-governmental Organisation
OECD	Organisation for Economic Cooperation and Development
OMVS	Organisation pour la Mise en Valeur du fleuve Sénégal
OKACOM	Okavango River Commission
ORASECOM	Orange-Senqu River Commission
PNA	Parallel National Action
PPA	Power Purchase Agreement
RBO	River Basin Organisations
RIDMP	Regional Infrastructure Development Master Plan
RISDP	Regional Indicative Strategic Development Plan
RSAP 3	Regional Strategic Action Plan for Water 2010–2015
SASB	Sustainability Accounting Standards Board
SADC	Southern African Development Community
SADCC	Southern African Development Coordinating Conference
SAHPC	Southern African Hydropolitical Complex
SAPP	Southern African Power Pool
SARTAC	Southern African Regional Technical Advisory Committee
SEEA	System of Environmental-Economic Accounting
SEEAW	System of Environmental-Economic Accounts for Water

SEFA	Sustainable Energy For All
SPI	Social Progress Index
SPV	Special Purpose Vehicle
SWI	Shared Watercourse Agreements and Institutions
UNEP	United National Environment Program
UN-DESA	United Nations Division of Economic and Social Affairs
VW	Virtual Water
WBCSD	World Business Council for Sustainable Development
WCI	Water Crowding Index
WEF	World Economic Forum
WEF Nexus	Water–Energy–Food Nexus
Westcor	Western Corridor
WFD	European Union—Water Framework Directive
WRTC	Water Resource Technical Committee
WSP	Water Sector Plan
WSRG	Water Strategy Reference Group
WWF	World Wildlife Fund
ZAMCOM	Zambezi Watercourse Commission
ZESA	Zimbabwe Electricity Supply Authority
ZIZABONA	Zimbabwe/Zambia/Botswana/Namibia

Introduction

Anna Entholzner

Abstract This chapter provides an overview of the development agenda of the Southern African Development Community and elaborates on the potential impacts of climate as well as demographic and economic change.

1 The Southern African Development Community

The Southern African Development Community (SADC) consists of 15 member states and was created as a response to the need to align development aid and to coordinate the struggle against colonialism and racism. After the fall of the South African apartheid regime, the focus of the community shifted and was formalised with the ratification of the SADC Treaty on 30 September 1993.

As stated in Article 5 of the SADC Treaty, the community's main objectives are to:

- achieve development, peace and security, and economic growth, to alleviate poverty, enhance the standard and quality of life of the people of Southern Africa, and support the socially disadvantaged through regional integration;
- promote self-sustaining development on the basis of collective self-reliance, and the interdependence of member states;
- achieve sustainable utilisation of natural resources and effective protection of the environment.

A. Entholzner (✉)
Climate Resilient Infrastructure Development Facility, Pretoria, South Africa
e-mail: anna.entholzner@cridf.com

© Springer International Publishing Switzerland 2016
A. Entholzner and C. Reeve (eds.), *Building Climate Resilience through Virtual Water and Nexus Thinking in the Southern African Development Community*, Springer Water, DOI 10.1007/978-3-319-28464-4_1

2 The Ambitious Agenda of SADC

The vision of SADC is one of a Common Future, a future within a regional community that will ensure economic well-being, improvement of the standards of living and quality of life, freedom and social justice as well as peace and security for the people of Southern Africa. This vision is achieved largely through the Regional Indicative Strategic Development Plan (RISDP) (SADC 2004) which is outlined as a comprehensive development and implementation framework for regional integration over a period of 15 years (2005–2020). It is designed to provide clear strategic direction with respect to SADC programmes, projects and activities. One of the key pillars of the RISDP lies in the activities of the Infrastructure and Services Directorate whose responsibilities cover coordinated infrastructure planning in a number of sectors including energy, transport, tourism and, water resources management and development. Signed at the SADC Summit in August 2012, the SADC Regional Infrastructure Development Master Plan (RIDMP SADC 2012) guides development in the key infrastructure sectors of water, energy, transport, tourism, meteorology and telecommunications. The SADC RIDMP acts as a framework for planning and cooperation with development partners and the private sector. There is a separate RIDMP document for each sector and the master plan will be implemented over three five-year intervals—short term (2012–2017), medium term (2017–2022) and long term (2022–2027). This is in line with the SADC Vision 2027, a 15-year implementation horizon for forecasting infrastructure requirements in the region.

SADC and the member states do not accept the current level of development and associated opportunities, equity and standards of living. They have set very ambitious goals for the 'Vision 2027' targets, drawing on the full range infrastructure needs from all key sectors. In most cases, SADC have cited global benchmarks; a clear sign that they are ambitious on behalf of their citizens. Whereas the sub-set of targets that are associated with water, and water related infrastructure are set out in Table 1, the current status of key water-related metrics in SADC in outlined in Table 2.

The current situation, especially in relation to key 'foundation' enabling infrastructure (e.g. water and sanitation service provision, clean energy service provision, food security, etc.) underlines the magnitude of the task. SADC countries currently have significant deficits in provision of basic services to existing populations. The deficit in the case of water provision is 39 % (more than 100 million people) and in the case of sanitation it is 61 % (nearly 160 million people). This, combined with significant population growth rates—five SADC countries have population growth rates over 3 %, with an average growth rate of 2.6 % for all SADC countries in the last decade (United Nations 2010–2015) will place enormous burdens on the region's economies. By 2027, at 2.6 % growth rate, there will be approx. 350 million people—close to another 100 million compared to 2015.

Table 1 Vision 2027 targets for water-related infrastructure

Sector	Vision 2027 targets
Surface water storage	25 % of Actual Renewable Water Resources (ARWR) stored, to meet SADC regional demand. Eventual target is 75 % stored as best practice is 70–90 % of ARWR stored
Agriculture	10 million hectares of irrigation (20 % of potential). World average is 20 %
Hydropower	75 GW installed to meet SAPP targets and exports to other Renewable Energy Companies (REC). This is 50 % of the total potential
Water supply	75 % of 350 million people served. Eventual target is 100 % served
Sanitation	75 % of 350 million people served. Eventual target is 100 % served
Water abstraction	264 km^3/year abstracted to meet an expected increase in water demand

Table 2 Current status of key water-related metrics in SADC

Sector	Current status
Surface water storage	14 % of Annual Renewable Water Resources (ARWR) stored (includes Kariba and Cahora Bassa dams)
Agriculture	3.4 million hectares (7 % of potential) irrigated
Hydropower	12 GW (8 % of potential) installed
Water supply	61 % of 260 million people served
Sanitation	39 % of 260 million people served
Water abstraction	44 km^3/year abstracted

3 Meeting the 2027 Vision

As indicated in Table 3 the investment requirements to achieve the infrastructure deficits are massive. The scale of resources required—financial as well as human, organisational and institutional, are simply enormous and are orders of magnitude greater than the resources that have been historically available. SADC needs to first 'catch up' on the provision of basic services and then accelerate to try and achieve their ambitious 2027 targets.

Table 3 The gap to be filled

Sector	Gap
Surface water storage	An additional 11 % of ARWR to be stored
Agriculture	An additional 6.6 million hectares to be irrigated
Hydropower	An additional 63 GW to be installed
Water supply	An additional 14 % of 350 million people to be served
Sanitation	An additional 36 % of 350 million people to be served
Water abstraction	An increase to 220 km^3/year abstracted

4   Extraneous Factors

Many extraneous factors impact on the ambitions of SADC; the key ones are indicated below.

4.1   Climate Change, Rainfall Patterns and Water Use

As the region is characterised by a wetter North and a drier South, shifting rainfall patterns, changing runoff and increased evaporation as a result of climate change are likely to change both the availability of and increase water demands. Various climate change models confirm the following climate change patterns within the SADC region (see Fig. 1). Overall, climate change is expected to be an exacerbating factor for regional water stress, increasing variability across time and space.

4.2   Demographic and Economic Change

In 2011, population growth rates in SADC were 2.68 % with the SADC mainland states having a total population of approx. 260 million (in 2013) and the aggregated Gross Domestic Product (GDP) growth rate was 5.14 %, widely varying 1.3 % (Swaziland) to 7.3 % (Mozambique). Three of SADC's member states (Angola, Mozambique and Zambia) are listed in the world's ten fastest growing economies. While South Africa's GDP growth rate is at 3 %, the relative size of its economy, which is nearly twice the rest of SADC combined means it represents a greater proportion of the aggregated SADC GDP growth and hence increased demands for water, food and energy.

Fig. 1 Shifting climate and rainfall patterns

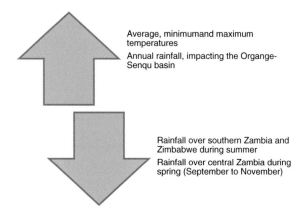

Average, minimumand maximum temperatures

Annual rainfall, impacting the Organge-Senqu basin

Rainfall over southern Zambia and Zimbabwe during summer

Rainfall over central Zambia during spring (September to November)

SADC is one of the world's fastest growing regions, characterised by a rapidly growing urban middle class, an increasing gap between wealthy and poor[1], and a Human Development Index below the global average of 0.66. The low income status of most of the countries means that they are particularly vulnerable to climate change as they heavily rely on agriculture. The agricultural sector contributes between 4 and 27 % to the region's GDP and 13 % of overall export earnings are from agriculture, while 70 % of the population in SADC depends on agriculture for food, income and employment. Regionally, the Democratic Republic of the Congo (DRC), Malawi, Mozambique, Zambia and Zimbabwe have a significant portion of their economy and labour linked to agriculture. These countries are also typified as being the most vulnerable to climate change.

Energy demand in SADC is growing at greater than 5 % and the availability of energy is a significant driver of regional economic growth. All the SADC states are actively pursuing increased energy production—through hydropower, thermal energy or the exploitation of shale gas reserves.

5 An Alternative Approach: Virtual Water and the Water-Food-Energy Nexus

Virtual Water represents water that is embedded in crops, livestock, and industrial items and services; it reflects the total amount of water used to produce goods and services. The basic concept of Virtual Water was developed in the early 1990s (Allan 1998), but has only been adopted relatively recently by the international community as a component of the analysis of water security. The transfers of Virtual Water in traded products can play a critical role in determining water security, especially where regions experience a range in water resource stress or water scarcity.

Virtual Water has blue, green and grey water components. Blue water is that water abstracted from surface or groundwater resources used to produce the goods; green water is the water derived from rainfall; whereas grey water is the volume of water that would be required to restore the water quality impact of any blue water abstracted to produce the goods, or dilute any return flows. The water footprint is the total volume as well as the proportions of blue, green and grey water embedded in the product (Allan 2011).

One of SADCs most defining characteristics with respect to Virtual Water and nexus thinking is the variability in water availability—geographically and temporary. Whereas the southern basins face physical water stress (see Chapter "The Future of SADC: An Investigation into the Non-political Drivers of Change and Regional Integration") while supporting the region's strongest economies and limited options

---

[1]The countries in the region some of the world's highest Gini-Coefficients (world average is 39 %), ranging from Tanzania (37 %) to Namibia (64 %) and South Africa (63 %).

for further water resource development through storage, the northern basins are better watered but food production does not as yet support strong economies in these basins, and water storage infrastructure is mostly limited to large hydropower schemes.

SADC as a whole is a net importer of Virtual Water in agricultural goods, importing 43 km^3/a, and exporting 25 km^3/a. However, of all the SADC agricultural exports, only around 40 % is traded within the region and trades are dominated by South Africa, which makes up 45 % of the total imports, and 56 % of the total exports. The significant Virtual Water importers are Angola (imports 98 % more Virtual Water than it exports), followed by the DRC (83 %), and Zimbabwe (71 %)—all of which are considered to be water rich countries. The significant exporters are Swaziland which exports 851 % more than it imports, followed by Malawi (495 %) and Zambia (250 %).

The nexus derives from the interaction, competition and trade-offs required between food and energy production as blue water users. A regional perspective on the nexus would be to produce the food and energy in places where competition for dwindling water resources would be lowest or to consider the full water footprint available for production when assessing potential trade-offs. Trading in goods with embedded water or making better use of the full water footprint may offer alternatives to physical transfers or storage at lower costs and environmental impacts while fostering regional cooperation and net benefits. Initial analyses suggests that sovereign security could be underpinned by regional water, food and energy security; reducing the impacts of local water shortages and contributing to building regional climate resilience.

Virtual Water and nexus perspectives may therefore offer a different view for national and regional infrastructure planners. Alternatives to large scale regional North to South water transfer infrastructure, could make better use of the total water footprint and Virtual Water trades in food and electricity, promoting regional integration and net benefits for the whole region, without threatening sovereign security. These concepts may also highlight regional benefits to be gained from infrastructure primarily aimed at satisfying national needs, and an alternative take on negotiating the reasonable and equitable use of water in shared watercourses.

The hypothesis is that Virtual Water and nexus thinking might provide an alternative approach for regional development planners. The following Chapters address the complex set of metrics behind implementing this challenge as shifting thinking from water, food and energy self-sufficiency to sovereign security through regional strategic planning. There are clearly considerable political, social and economic challenges to introducing these concepts at both national and regional levels. The Climate Resilient Infrastructure Development Facility, consequently, established a panel of experts with considerable experience in regional and global development, water and Virtual Water concepts. This volume represents the opinions of this panel aimed at highlighting key considerations for taking these concepts further in meaningful engagements with national and regional planners and is organised as follows:

Chapter "Quantifying Virtual Water Flows in the 12 Continental Countries of SADC", written by Dr David Philipps and Stephen Boyall provides a summary of the development of a Virtual Water Database and summarises some key findings around Virtual Water trade in agricultural products and electricity while also offering an overview of water accounting in the region. Prof Anthony Turton then postulates the future of SADC under the pressures of climate change and shifting water scarcity in Chapter "The Future of SADC: An Investigation into the Non-political Drivers of Change and Regional Integration", examining the non-political drivers and change and regional integration and the role Virtual Water can play. The following Chapter will then examine Virtual Water and the nexus concepts in the context of national and regional development planning, highlighting the key drivers of development. Barbara Schreiner then poses options to include Virtual Water, water footprinting and nexus thinking into water allocation processes both on a national and transboundary scale in Chapter "Mechanisms to Influence Water Allocations on a Regional or National Basis". The following Chapter "Electrical Power Planning in SADC and the Role of the Southern African Power Pool" outlines electrical power planning in SADC and the role of the Southern African Power Pool in introducing potential benefits of reduced stress, costs of generation and carbon emissions. Dr Guy Pegram and Dr Hannah Baleta examine the role of private sector water stewardship and water footprinting in managing water in Chapter "Virtual Water and the Private Sector". Prof. John Anthony Allan shows that there are a number of international conditions and trends, first in water, energy and food supply chains and secondly, in international trade and demography in his Chapter "The International Experience". Reginald M. Tekateka outlines the challenges of introducing the concepts of Virtual Water and the nexus within SADC in his concluding Chapter "Embedding the Virtual Water Concept in SADC". Dr Charles Reeve outlines additional ways of fostering collaboration and cooperating within SADC in the postscript to this book.

References

Allan JA (1998) Virtual water: a strategic resource. Global solutions to regional deficits. Groundwater 36(4):545–546

Allan JA (2011) Virtual water: tackling the threat to our planet's most precious resource. I.B. Tauris, London

SADC (2004) Regional Indicative Strategic Development Plan (RISDP)

SADC (2012) Regional Infrastructure Development Master Plan

United Nations 2010–15 available at http://esa.un.org/unpd/wpp/Download/Standard/Population/. Accessed 23 Mar 2016

Quantifying Virtual Water Flows in the 12 Continental Countries of SADC

David Phillips and Stephen Boyall

Abstract The Climate Resilient Infrastructure Development Facility (CRIDF) has developed a data platform for Virtual Water transfers amongst the 12 continental countries of the Southern African Development Community (SADC) and with the rest of the world, embedded in agricultural products and in electricity supplies. This, together with a Global Social and Development Index database, has formed the basis for many of the analyses in other Chapters of this book. This Chapter provides a summary of the processes and source data used in developing develop the databases, as well as some examples of the potential use of the data platform. The final section provides a brief summary of the techniques involved in economic accounting of water, and the current position of the continental SADC countries in that regard. Notably, economic accounting of water relates to blue water in isolation, but the technique nevertheless provides useful complementary information to that obtained through considerations of the rest of the water footprint and the economy behind Virtual Water transfers.

1 Introduction

The Climate Resilient Infrastructure Development Facility (CRIDF) is an initiative of the Department for International Development (DfID) of the United Kingdom Government. CRIDF seeks to develop climate resilience in poor communities through the construction of infrastructure in the continental countries of the Southern African Development Community (SADC), thereby promoting the peaceful management of

Dr. David Phillips, a freelance consultant—deceased

D. Phillips · S. Boyall (✉)
Adam Smith International, London, UK
e-mail: steve@steveboyall.com

© Springer International Publishing Switzerland 2016
A. Entholzner and C. Reeve (eds.), *Building Climate Resilience through Virtual Water and Nexus Thinking in the Southern African Development Community*, Springer Water, DOI 10.1007/978-3-319-28464-4_2

shared waters. However, CRIDF has recognised that in the longer term, peaceful cooperation in the increasingly water-stressed SADC region also requires attention to larger strategic infrastructure investment and planning. This wider focus is inherently pro-poor, protecting access to water for community-based small-scale infrastructure, but also supporting shared economic growth by planning water-related investments which recognise the potential for Virtual Water trading as a means of addressing regional water, food and energy security.

Virtual Water represents water that is 'embedded' in crops, livestock, and industrial items and services, having been used to produce these. The basic concept of Virtual Water was developed in the early 1990s (Allan 1998), but has only been adopted relatively recently by the international community as a component of the analysis of water security. The transfers of Virtual Water in traded products can play a critical role in determining water security, especially where regions (or trading partners) display a range in water resource stress or water scarcity. The continental SADC countries exhibit a wide range in water resource availability, the southern portions of the African continent being much more water-stressed than the northern SADC states. Changes to international trading patterns could help to alleviate such stress, and can sometimes be mutually beneficial to partner countries —in certain instances, reducing the expenditure and infrastructure required to develop additional blue water resources,[1] and building regional climate resilience. From one perspective, agricultural production and international trade patterns are driven by a wide range of factors, many of which are intractable and not easily changed or influenced. However, large multi-national companies present in most or all of the SADC countries may minimise their climate change risk by considering Virtual Water in their supply chains. In any event, sound and politically aware arguments will be needed to underpin any shifts in attitudes towards regional rather than sovereign water and energy security, and climate resilience.

This chapter outlines the data assembled on Virtual Water transfers in agricultural products and in electrical supplies traded by the continental SADC countries (with each other and in some cases with the rest of the world), and has been created to act as a partial basis for further work with the intention to help the development of sound arguments and an evidence base grounded in improved resource management and shared economic benefits.

---

[1]Blue water is present in surface waters and aquifers, and is the classical focus of studies on the hydrosphere. However, green water (known sometimes as soil moisture) is also of great importance in the agricultural sector in particular, and grey water (volumes required to account for polluting effects) can also be of significance.

2 Data Sources: Quality Assurance/Quality Control

2.1 Data Relating to Agricultural Products

As noted in greater detail in the following section, the Virtual Water components and transfers relating to all internationally traded agricultural products (crops and livestock) for all of the 12 continental countries within SADC are identified.

In most cases, the resources available precluded the collation of primary data from specific agricultural sectors, or the data available from SADC states which are not accessible through the internet. The primary databases employed were therefore mainly of a global nature, and included:

- Chapagain and Hoekstra (2004) for bulk Virtual Water transfers;
- Mekonnen and Hoekstra (2011) for water footprints;
- Statistics from the Water Footprint Network (http://www.waterfootprint.org);
- Import and export data from Trade Map (http://www.trademap.org/), which rely on statistics from UN Comtrade (http://comtrade.un.org/); and
- National and regional data for South Africa from the Department of Agriculture, Forestry and Fisheries (http://www.daff.gov.za/), supplemented in particular scenarios by other national/regional data.

The first four sources cited above are those utilised by most or all researchers on Virtual Water internationally, and are widely recognised as offering state-of-the-art information that has been comprehensively quality-assured by the respective authors. There is nevertheless some controversy concerning detailed levels of water footprint data, and it is cautioned that some authorities suggest that such detailed water footprint estimates should be treated with caution. The current authors have noted that such caution is merited, on occasion.

The Trade Map and UN Comtrade data include some inconsistencies which have been noted by the project team. These involve occasional mismatches between the export statistics for a specific product involving a particular country of destination and time period, and the mirror-image import statistics of the country of importation for the same product and time period. In order to complete a comprehensive Quality Assurance check in line with best practice, the following advice from UN Comtrade has been followed:

> UN Comtrade disclaimer: "Imports reported by one country do not coincide with exports reported by its trading partner". (Point 5)

In such circumstances, UN Comtrade recommends the use of data for imports when evaluating trade patterns, as such data are independently verified through the customs protocol of a receiving country. The Virtual Water trade database for agricultural products has therefore been constructed in this way to ensure that the most accurate and verifiable data sets have been used (UN Statistics 2014).

The national statistics for South Africa were derived directly from the primary governmental source of such information, and this is also the case for other specific state-centric data used in the examples quoted in the present report. Trade-related data for various countries have been compared in general terms to information in the FAOSTAT database of the Food and Agricultural Organisation, and this acted as a further quality assurance check.

The data used to compile the agriculturally-related database for the 12 continental SADC states are of a country-specific nature and are hence broad in scope. While the specific treatment of these data adds considerable value and creates a data platform to support the development of regional water and food scenarios, scope remains for more detailed analyses within specific sectors.

At the present stage of the work, data for the trade in electricity amongst the SADC countries have been provided primarily by Eskom and by the staff of the SAPP offices in Harare, Zimbabwe.[2] Further information of relevance (and to support the primary sources) was accessed from online sources of the US Energy Information Administration, which is an entity under the United States Department of Energy based in Washington D.C. Various published sources were employed to check the general patterns of trade in electricity which are subject to a degree of change over time (Economic Consulting Associates 2009a, b; South African Power Pool 2009; Southern Africa Regional Integration Strategy Paper 2011).

Data reported here on the consumptive use of water in the generation of electricity and on the renewable water resources of specific countries were abstracted primarily from three sources:

- National statistics for South Africa, compiled by Eskom (see footnote [2] below);
- The report of Beilfuss (2012) for hydroelectric facilities; and
- The FAOSTAT database of the Food and Agricultural Organisation[3].

Cross-checks on these data were completed using published information for the consumptive use of water by electricity generating facilities elsewhere in the world (Torcellini et al. 2003; World Bank 2010). There is considerable variation in the methodologies used internationally to calculate consumptive water use in the generation of electricity. The current study does not seek to provide highly precise full life-cycle estimates of consumptive water use at each specific facility of relevance in the 12 continental countries of SADC. Such precise estimates are in any event not of utility in the context of this work, which relates to the introduction of Virtual Water and nexus-based thinking into the regional debate on water resource management and development. Hence, the estimates of consumptive water use in the generation of electricity as shown here are of an indicative nature, and it is argued that this is more than sufficient to act as the backdrop for future discussions of preferred approaches to electricity generation in the future in the region.

Data on economic accounting of water were amassed from a number of sources internationally and within SADC.

---

[2]Data from Eskom were provided by Dr. Dave Lucas, and those from SAPP were provided by Dr. Lawrence Musaba.

[3]FAOSTAT database available at http://www.fao.org/statistics/en/.

3   Virtual Water Transfers in Agricultural Products

3.1   The Development and General Structure of the Database

The data platform that has been established provides information at three distinct tiers of detail which are inter-related, as shown in Fig. 1. These are described below, with examples of their use also being provided.

- The top-tier data are of a generic nature, involving net and gross estimates at country level of all internationally traded crops and livestock (separately in those two categories). External trading parties are identified specifically within the continental SADC group of 12 states, while trade with the rest of the world is cited as a single entity in the main platform. These data characterise international trade patterns in a broad manner, showing whether specific countries are net importers or exporters of Virtual Water. With more detailed interrogation, the data platform can reveal 'water savings' either for particular countries or globally, given the existing trade pattern or following pre-supposed changes to the existing pattern. Trade with the rest of the world can also be broken down into imports/exports involving specific individual countries or groups of countries, where this is of interest.
- The middle tier of detail provides data on specific traded items (including all crops and livestock subject to international trade involving SADC countries), as either single products or small groups of products according to the citations in the available trade data (as 4-digit categories of the Harmonised System, these

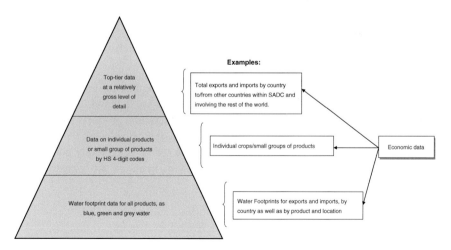

Fig. 1 Diagram of the types of data available for the trade in agricultural products amongst the continental SADC countries

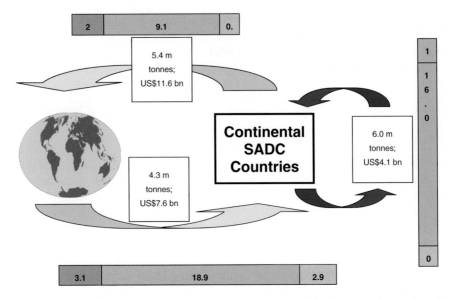

Fig. 2 Imports and exports of agricultural products (in total) in 2012 amongst the continental SADC countries, and between these and the rest of the world (Tonnages and values are shown in text boxes; the accompanying Virtual Water transfers are shown by 'colour' in cubic kilometres)

numbering 117 products in total in the present context[4]). Data are available in this tier of detail as country-wide exports and imports, with tonnages and total Virtual Water equivalents being provided. Regional data (within countries) can also sometimes be accessed.

- The lower tier of detail superimposes water footprints onto each product and country pairing, with blue, green and grey water all being identified in full. While this averages the footprint information across single countries, it provides a useful starting point for more detailed analysis, which can again involve regional differences where these are of interest.

The tiers of the data platform inter-relate, in the sense that information derived from one tier can be related to that in any other tier. Figure 2 shows an example of this, and also provides an overview of the trade in agricultural products between the continental SADC countries and the rest of the world. It is notable that the overview shown in Fig. 2 suggests that the trade pattern as a whole is generally coherent in relation to the continental SADC states, at least in terms of the water-stressed countries.

---

[4]The Harmonised System—also known as the Harmonised Commodity Description and Coding System—was introduced in 1988 and is used by most countries in the world to characterise trade. It is run and maintained by the World Customs Organisation based in Brussels, which has over 170 members.

Thus, the 12 continental SADC countries import agricultural products of lower tonnage and lower value from the rest of the world than they export thereto, but the imported products have about twice the Virtual Water content compared to the exports. The trade in agricultural products amongst the continental SADC countries is of a generally similar order of magnitude to the external trade, but of lower total value—and this is also accompanied by significant Virtual Water transfers. Greater levels of detail within the data platform reveal more nuanced patterns, however, not all of which appear so strategically coherent. Thus, for example, the exports from the continental SADC countries to the rest of the world have a higher percentage of blue water, compared to the imports (see Fig. 3)—and this would appear at first sight to be contrary to the interests of the water-stressed countries within the SADC group of states.

It is also notable that agricultural products imported into SADC from the rest of the world tend to have a proportionally larger grey water content than those derived from and traded internally within SADC. The water footprint of agricultural products in the SADC countries contains only a minor grey water component, although the proportions of the water footprint taken up by blue water and green water vary considerably for some products, from country to country and also from place to place within a single country. Certain examples which follow here have ignored the grey water component of the overall water footprint for commodities traded within SADC, for simplicity.

As shown in Fig. 1, the economic values of agricultural products have also been captured in the creation of the data platform, these being abstracted from the Trade

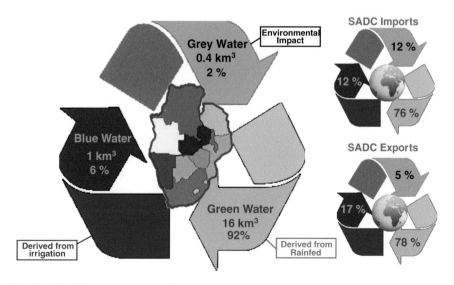

Fig. 3 Virtual Water components of imports and exports of agricultural products (in total) in 2012 amongst the continental SADC countries, and between these and the rest of the world (Virtual Water transfers are shown by 'colour' in cubic kilometres and also in percentage terms in relation to trade with the rest of the world)

Map and UN Comtrade statistics. The economic data can be utilised in a variety of different fashions, e.g. to calculate standard 'water productivity' values (US dollars generated per unit of blue water). A reliance on green water or grey water has a significant effect on the standard 'water productivity' values, and 'Virtual Water productivity' is proposed as a more useful indicator. It is also notable that the Virtual Water productivity of specific crops has been found to vary widely from one location to another within southern Africa, this being ascribed to a combination of the efficiency of irrigation, the yields achieved in crop production, and the prices paid for the exported crops. Such variations are of obvious significance in terms of the potential for poverty reduction in specific circumstances, but the caveats noted previously in terms of the robustness of water footprint data at high levels of detail should be kept in mind during such analyses. These regional differences notwithstanding, the Virtual Water productivity of SADC imports from the rest of the world is approximately U\$ $0.3/m^3$, while that of SADC exports to the rest of the world is U\$ $1.00/m^3$ (see Fig. 2).

The complete agricultural dataset is very substantial, the file size in Excel being almost 18 megabytes; the database can be requested from CRIDF. The data are made available for each of the 12 SADC countries addressed by the work as a whole, as well as individually. This assists users of the database to access information relating to their own specific interest. The units employed have been chosen to create sufficient precision, while avoiding the use of decimal points (i.e. a rounding up/down method has been utilised). This rounding up/down technique occasionally results in minor differences between information synthesised from the data platform, and that published by Mekonnen and Hoekstra, but such distinctions are not of any significance.

4   Examples of the Use of the Agricultural Data Platform

Given the very considerable breadth of the data platform and its inherent detail, many examples could be provided of its use relating to traded agricultural commodities. However, in the interests of simplicity and brevity, this Chapter is restricted to only a few of these.

4.1   Examples of Top-Tier Data

Table 1 shows data for the net transfers of Virtual Water in the imports and exports of crops, livestock and industrial products amongst the 12 continental SADC countries.

Five of the 12 countries are net exporters of Virtual Water in traded products (Malawi, Mozambique, Tanzania, Zambia and Zimbabwe). However, it is cautioned that these high-level data are derived from Chapagain and Hoekstra (2004)

Table 1 Net Virtual Water imports/exports internationally for the 12 continental countries of SADC

Country	Net virtual water imports (MCM/year)			
	Crop products	Livestock products	Industrial products	Total trade
Angola	206	447	140	793
Botswana	376	−62	54	369
DRC	136	107	59	302
Lesotho	ND	ND	ND	ND
Malawi	−646	8	32	−607
Mozambique	−1112	−6	54	−1064
Namibia	45	−96	88	37
South Africa	1426	−293	1011	2145
Swaziland	134	41	41	216
Tanzania	−2203	−41	83	−2161
Zambia	−271	−14	−38	−323
Zimbabwe	−3032	−319	103	−3247

Data from Chapagain and Hoekstra (2004). *ND* No data

and refer to the late 1990s and early 2000s. Trade patterns for agricultural products change somewhat over time, and more recent data reveal subtle differences to the pattern shown in Table 1, these being explained below.

To break the net data down and add a layer of detail (albeit still at the top tier of the pyramid as presented in Fig. 1), Figs. 4 and 5 show country data for imports and exports of agricultural products (crops and livestock together) during the year 2012 by the individual SADC countries, and the respective values of those traded products (as thousands of US dollars).

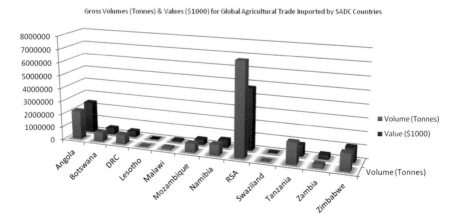

Fig. 4 Imports of agricultural products by the continental SADC countries, and their total values

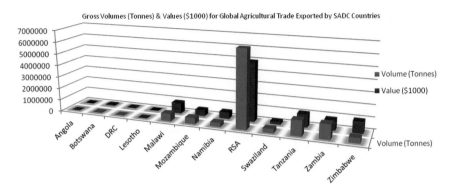

Fig. 5 Exports of agricultural products by the continental SADC countries, and their total values

South Africa dominates the overall profile for traded agricultural products in SADC, with high volumes for both imports and exports. The tonnages of imports outstrip those of exports (as is the case for the Virtual Water transfers shown in Table 1), but the economic values show a reverse trend, with exports being of greater total value than imports. In general terms (and as would be expected), the Virtual Water transfers shown in Table 1 align well with the data on traded tonnages and values in Figs. 4 and 5.

A number of the continental SADC countries are presently significant net importers of agricultural products (and Virtual Water), these including Angola, Botswana, the DRC, Tanzania, and Zimbabwe. Mismatches appear between the gross Virtual Water data shown in Table 1 and the data in Figs. 4 and 5, for both Tanzania and Zimbabwe. These relate to the distinct ages of the datasets, the annually averaged data in Table 1 referring to a five-year period in the late 1990s and the early 2000s, whilst the information in Figs. 4 and 5 pertains to the year 2012.

Notably, Malawi, Swaziland and Zambia are presently net exporters of Virtual Water, and are also substantial net earners of foreign exchange in agricultural products. While South Africa is a net foreign exchange earner in agricultural products, it remains a net importer of Virtual Water. However, when data are contextualised against the size of the economy, Malawi, Swaziland and Zambia stand out as substantial exporters of agricultural products in terms of both Virtual Water transfers and value—perhaps making these countries particularly vulnerable to reduced rainfall and run-off due to climate change.

4.2   Examples of Middle-Tier Data

As noted in Fig. 1, the 'middle tier' of the data platform is populated by product-related data the precise products included being listed in Annex 1. Certain

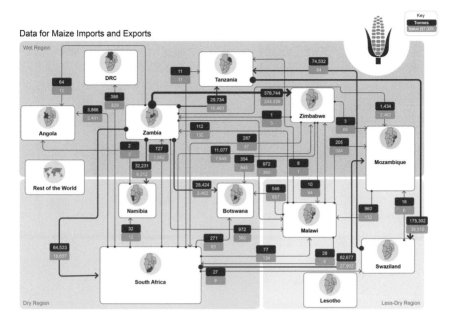

Fig. 6 Import and export of maize within SADC, in terms of tonnages and respective values (Trade data with the rest of the world are not shown)

products stand out amongst the internationally-traded commodities in SADC, and maize is one of these—and is also of course important as a staple foodstuff, being produced in all of the SADC countries to assist in feeding their respective national populations.[5]

Figure 6 Shows the tonnages and values of maize imports and exports amongst the SADC countries only. It is noted that data involving trade with the rest of the world are not shown, and are minor by comparison in any event. The trade pattern within SADC is complex, and maize represents a product of significant interest to the present studies as a result (see also the data on water footprints for traded maize, in the following sub-section).

The detailed trade patterns for ground nuts and sugar provide examples of three very different agricultural products. The tonnages and values of ground nuts and sugar traded within SADC are shown at Figs. 7 and 8, respectively. Information of this type can be provided from the data platform for any of the single products or groups of products.

---

[5]It is notable in passing here that the data platform created concentrates on internationally traded items, as possible changes to the Virtual Water transfers internationally are the primary issue of interest. However, countries can also free up blue water resources by altering their patterns of crop and livestock production to feed their own national population, and this represents an area which could well be of major interest to the Virtual Water studies in southern Africa in the future.

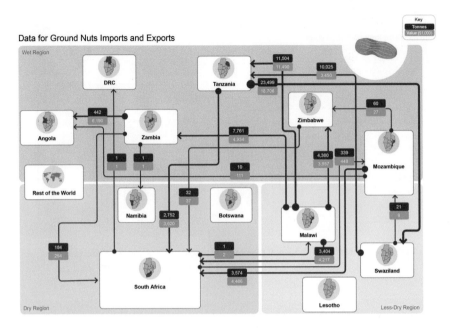

Fig. 7 Import and export of ground nuts within SADC, in terms of tonnages and respective values (Trade data with the rest of the world are not shown)

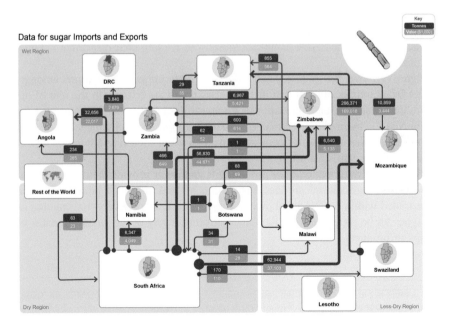

Fig. 8 Import and export of sugar within SADC, in terms of tonnages and respective values (Trade data with the rest of the world are not shown)

4.3   Examples of Lower-Tier Data

Lower-tier data relate to water footprints, which can be generated for trade patterns as a whole as shown in Fig. 2, or for individual products (or groups thereof). Figures 9, 10 and 11 show data on the blue and green water transfers between the SADC countries allied to the trade in maize, ground nuts and sugar. These graphics provide the key Virtual Water components to complement the data on tonnages and values shown in Figs. 6, 7 and 8.

Interesting patterns are revealed when the water footprint data are interrogated in detail. Examples of these are shown in Text Boxes 1 and 2, which address water productivity values of various types, and crop yields. As noted in Text Box 1, blue water productivity is generally quite low for sugar production in southern Africa, averaging US$2.78/m^3. Higher blue water productivity is noted for groundnuts (mean of US$4.41/m^3), with very much higher values for maize (averaging US $177/m^3). These figures reflect the reliance on blue water (and irrigation) for the crops involved. It is noted that while the data on water productivity are to some extent dependent on exactly where the crop is grown (and may be susceptible to the issues described previously in terms of the robustness of detailed water footprint information), average data have been used here.

When analysed on the basis of the full Virtual Water footprint, productivities average only US$0.38/m^3 for groundnuts, but somewhat higher for sugar

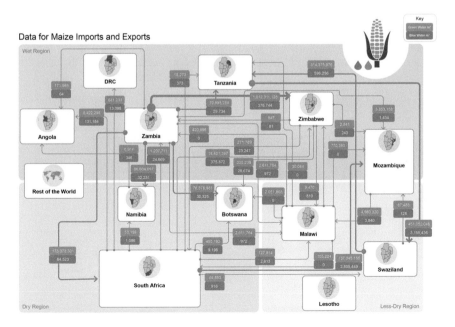

Fig. 9   Import and export of green and blue water in maize traded within SADC (Trade data with the rest of the world are not shown)

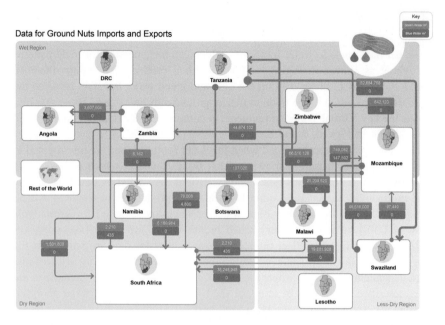

Fig. 10 Import and export of green and blue water in ground nuts traded within SADC (Trade data with the rest of the world are not shown)

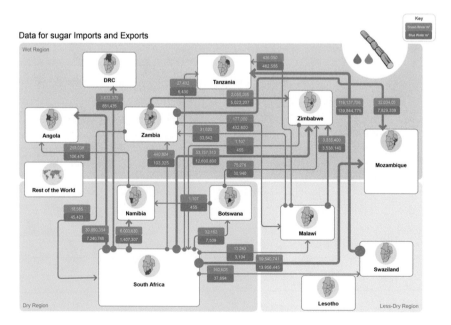

Fig. 11 Import and export of green and blue water in sugar traded within SADC (Trade data with the rest of the world are not shown)

(US$0.69/m^3) and considerably greater for maize (US$1.28/m^3). The most striking facet of the data relates to the very large differences between the water productivities calculated for specific sources/destinations of traded crops. It is clear from the relatively simple analysis completed to date that further focus on the patterns of crop production would materially affect both agricultural efficiency and economic returns, with potentially large effects on poverty in certain rural areas.

The second example provided here involves the production of sugar on a regional basis in both South Africa and Zimbabwe. In South Africa, sugar is grown primarily in KwaZulu Natal and Mpumalanga Provinces. Very considerable differences exist between these two locations in terms of the extent of irrigation of the crop, this being far greater in KwaZulu Natal (at 90 % of the total area devoted to sugar) than in Mpumalanga (only 14 % of the total area used for sugar production being irrigated). As expected, crop yields are lower for the non-irrigated product (56 tonnes/ha as an average in KwaZulu Natal), and rise considerably when irrigation is made available (89 tonnes/ha in Mpumalanga). In Zimbabwe, all sugar is grown in the Save River Basin, and this is all irrigated. The primary distinction relating to sugar production in this instance involves the average yields attained by subsistence farmers on small-holdings (34 tonnes/ha) and those achieved by commercial operations (76 tonnes/ha). Text Box 2 includes some initial analysis in this regard.

The examples cited above show that the data platform is of exceptional utility, and is capable of providing information at many levels of detail and specificity. It is important to note that the tool that has been created by the current work is not simply a cut-down version of the dataset published by Mekonnen and Hoekstra, but has been designed specifically for use in southern Africa to assist SADC in moving towards regional water and food security, creating a vision of water as a regional public good.

Text Box 1: An Example Concerning Virtual Water Transfers In Agricultural Products Traded Internationally in Southern Africa
'Water productivity' is considered by many commentators to be an important measure of the efficiency of agricultural activities. Generally measured as the financial return per volume of water used (US$/m^3), water productivity values as reported to date in the literature refer to a basis of blue water only. Where crops are grown with a heavy reliance on green water (and only limited blue water use), this increases the apparent water productivity.

The data platform was used to derive the standard water productivity values (based on blue water alone) and also for a new parameter, which could be termed the 'Virtual Water productivity'. The latter values are based on all forms of water used to grow a primary crop: blue, green and grey water, in combination.

Three distinct types of crops were selected to create this example, and the resulting data are summarised as follows:

- Average (blue) water productivity values were US$2.78/m^3 for sugar; US $4.41/m^3 for groundnuts; and US$177/m^3 for maize.
- Mean values for the Virtual Water productivity were US$0.38/m^3 for groundnuts; US$0.69/m^3 for sugar; and US$1.28/m^3 for maize.

The water productivity data show the influence of the blue water/green water mix used for each crop, in a very clear fashion. Virtual Water productivity figures provide a much more coherent base to compare the financial output from each crop, but effectively value each form of water equally.

The data for these three crops traded in various fashions between the SADC countries show great differences from place to place (trade to trade) in both water productivity and Virtual Water productivity—the averages cited above masking very large variations. Such information is not usually used by farmers or by governmental authorities in reaching decisions on preferred crops to be produced in specific locations.

Should public or private entities incorporate considerations of Virtual Water (or at least, the distinct types of water) into their planning in terms of which crop to grow, where?

Are governmental bodies the most appropriate to lead in any such an intervention, or would private sector farming interests be better engaged?

How likely is it that the SADC countries would agree to grow 'thirsty crops' in the north of SADC, and 'less thirsty crops' in the more arid southern regions, with trade between these addressing local demand for all the various types of crops?

Text Box 2: The Production of Sugar in Southern Africa

Yields for the production of sugar in South Africa and Zimbabwe are shown below:

Location	Type of production	Yield (Tonnes/ha)
KwaZulu Natal (RSA)	Dry land (not irrigated)	56.0
	Irrigated	70.7
Mpumalanga (RSA)	Dry land (not irrigated)	63.6
	Irrigated	89.5
Zimbabwe	Subsistence (small-holdings)	34.0
Zimbabwe	Commercial	76.0

As would be anticipated, the yields increase perceptibly when irrigation is made available (in South Africa), although the uplift in yield provided by irrigation is not particularly great (26 % in KwaZulu Natal, and 41 % in Mpumalanga). In Zimbabwe, the key determinant of yields of sugar involves the distinction between those attained by subsistence farmers on small-holdings, and the much higher yields achieved by commercial operations.

What level of government intervention may be countenanced, to attempt to improve crop yields in such scenarios?

Is the irrigation of sugar (which has low water productivity in general) a rational use of blue water supplies in the water-stressed areas of southern Africa? Should blue water be allocated to higher-value crops in such water-stressed areas, and how might this be achieved?

What forms of intervention may be promoted to increase subsistence-level yields of crops? Should subsistence farmers be encouraged to grow crops with higher water productivities, and how might this be achieved?

In the course of the construction of the data platform, the team noted that certain types of data could be added to the data platform. These include the following:

- Data on the national production and consumption of agricultural products (as opposed to internationally traded products), which would be of interest when parties wish to interrogate the efficiency of the sector in particular countries and provinces. The production data can be associated with areas farmed, yields, and economic values;
- Detailed data on the specific trade patterns of the SADC nations with individual countries in the 'Rest of the World' category, on a product-by-product basis or more generic platform, as may be desired.

In terms of high levels of detail, a few products do not have assigned water footprints, as no data are available internationally in this regard. Further work could also be completed to superimpose more detailed estimates of animal size on the data platform, to better assess the Virtual Water component of livestock products. However, given that this type of additional work is likely to require substantial resources, it should only be pursued on an as-needed basis.

5 Virtual Water Transfers in Traded Electricity

5.1 Consumptive Water Use in the Generation of Electricity

As noted in a previous section, there is considerable variation in the methodologies used internationally to measure the consumptive use of water in the generation of electricity. Some authors cite data based simply on the evaporative losses involved in the generation process, as this is generally by far the largest and most important component of overall net water use. By contrast, other authors attempt 'life cycle analyses' which extend to the source of the energy (and the process making this available), the equipment used, downstream transmission, etc. Such details are captured particularly clearly in a recent literature review by Meldrum et al. (2013).

The resources available to the current authors did not permit full life-cycle analyses of consumptive water use in the generation of electricity amongst all of the facilities in the continental SADC countries, and in any event these were not deemed to be required. Thus, CRIDF seeks to introduce the concepts of Virtual Water and the nexus into the general policy-related planning process within the region as a whole. While it is recognised that highly specific data on life cycle water use in certain scenarios might be of relevance in particular circumstances at a later time, the present report employs indicative data only, based on estimates of the evaporative losses of water in particular electricity generating technologies. Similarly, no attempt has been made to address the relatively minor sources of electricity within the continental SADC countries (wind, solar, etc.), and the focus here is on thermoelectric plants which dominate in the South of SADC, and hydropower facilities that are primarily located in the North of the region.

In thermoelectric power generation, the level of consumptive water use depends primarily on the selected method of cooling, with a significant difference being observed between once-through and recycling systems. By contrast, the consumptive use of water in the generation of electricity using hydropower varies greatly amongst distinct types of facilities, and also between specific plants of any one type.[6] Run-of-the-river systems which do not involve large impoundments have low consumptive water use, while facilities including large dams and reservoirs exhibit a much higher consumptive use of water. This is due to a number of factors, including in particular the surface area of any impoundment; the rate of evaporative loss from the reservoir; and the efficiency of the turbines. In the USA, Torcellini et al. (2003) cited a range for hydropower plants from close to zero consumptive use for run-of-the-river systems, to more than 208 L/kWh—very much greater at the higher end of the range than the consumptive water use in thermoelectric plants (see below).[7] The consumptive use of water by hydropower facilities in Africa can

---

[6]For generic data, see for example *Energy Demands on Water Resources. Report to Congress on the Interdependency of Energy and Water.* US Department of Energy, December 2006.

[7]It is notable that some of the water sources used to support thermoelectric power generation also involve impoundments, in some instances to increase assurance of supply. However, these are generally much smaller than those employed to generate hydropower, with minor evaporative losses—and they are also usually multi-use facilities, being employed in support of agricultural irrigation in particular. It is also arguable that the evaporation off the large hydropower-related impoundments such as Kariba and Cahora Bassa might not all be allocated to hydropower generation in isolation (where multiple use occurs), but in those cases the hydropower generation was the primary *raison d'etre* for the construction of the dams. The analysis provided here is thus considered to be generically robust, and in any event the very large distinction between consumptive water use in the two forms of electricity generation would persist, even where additional (minor) factors are taken into account.

be far higher than that of similar plants in the USA, due to the higher evaporative loss in many locations in Africa (although these also vary considerably from site to site). Importantly, increased evaporative loss due to climate change induced higher temperatures could increase net water use in hydropower. Given the higher than average temperature increases expected in the SADC region, climate change could play an increasingly important role in SADC's energy future.

Table 2 provides a brief summary of the energy sector in each of the 12 SADC countries addressed in this chapter, and this highlights two key points of relevance to Virtual Water transfers. South Africa generates about 80 % of the electricity produced in southern Africa as a whole, and this equates to approximately 40 % of the generation of electricity in the entire continent. Amongst the CRIDF countries addressed here, the northern states rely heavily on hydropower for the generation of electricity (see Table 3), while South Africa presently utilises coal-fired thermo-electric generation for the great majority of its electrical supplies, as shown in Table 4.

Recent data from Eskom reveal a consumptive use of water of 1.37 L/kWh in the generation of South African electricity, as a nation-wide average. This nation-wide average for the consumptive use of water in electricity generation in South Africa is competitive by comparison to performance elsewhere, e.g. Torcellini et al. (2003) reported an average consumptive use of 1.4–1.9 L/kWh for thermoelectric power generation by a large range of such facilities in the USA. It is notable that the newer thermal plants in South Africa are more water-efficient, and the gradual introduction of dry cooling will reduce water consumption even further.

Amongst the major hydropower plants in SADC (see Table 3), certain facilities in the Zambezi River basin are of particular significance due to their reliance on large dams which impound reservoirs of very considerable surface area—with highly significant evaporative losses. Data from Beilfuss (2012) were used by the CRIDF project team to determine the consumptive uses of water at the three major hydropower sites in the Zambezi River basin.

These revealed the lowest use at Itezhi-Tezhi/Kafue Gorge (64 L/kWh); intermediate values at Cahora Bassa (296 L/kWh); and by far the highest consumptive use of water at Kariba (1040 L/kWh). The particularly high losses at Kariba reflect the relatively shallow reservoir with a large surface area, and this is also demonstrated by data for electricity generation per reservoir surface area (0.3 MW/km^2 at Kariba, as opposed to 1.4 MW/km^2 for Cahora Bassa). The total evaporative losses at the hydropower facilities in the Zambezi River basin equate to a consumptive use of about 11 % of the mean annual flow of the system as a whole, and this creates major changes not only to the total water flow in the basin but also to the seasonal pattern of flows.[8]

---

[8]It could be argued that the construction of major dams for generating hydropower has effects not only on evaporation rates upstream, but also those downstream, due to the resulting changes in river flow dynamics/cross-section, etc. Such effects are certainly relevant in terms of basin-level water management, but they have not been taken into account here (and are believed in most cases to be minor, compared to evaporative losses at the reservoir sites).

Table 2 Overview of the energy resources and electricity supplies in each of the continental SADC countries

Country	Energy resources	Electricity supplies
Angola	Considerable oil and gas resources (global #18 in reserves) in ongoing development, these entirely dominating the national economy. Primary energy use still dominated by biomass, followed by oil, hydropower and natural gas	Approximately 70 % of internal generation from hydropower (Cuanza, Catumbela, Cunene Rivers). Gas-fired production likely to increase. Only 30 % coverage of the population currently. Not yet fully integrated into the SAPP, but a connection to Namibia is being established
Botswana	No oil or gas reserves. Moderate coal resource, essentially all utilized within the country	Significant importer of electricity, mostly from South Africa
DRC	Significant oil reserves, with more likely to be discovered. A minor oil exporter currently. Some coal and significant hydropower at Inga in particular. The development of Grand Inga could completely alter the African balance of power generation	Only a moderate net exporter of electricity, mostly from the Inga projects. New Power Purchase Agreement with South Africa relating to Inga III. Massive future potential as an exporter of hydropower from Grand Inga
Lesotho	No oil, gas or coal reserves; reliant on national biomass and imported fossil fuels	Minor generation through hydropower, including the Lesotho Highlands Water Project. Minor net importer of electricity from South Africa
Malawi	Very minor oil production at present. No gas currently exploited, but attempts are ongoing to develop oil and gas under Lake Malawi/Nyasa (involving a dispute with Tanzania)	Essentially all electrical power generated in-country mostly through hydropower—with significant new hydropower plans, and no significant imports or exports
Mozambique	No significant oil reserves. Natural gas exploited from the onshore Pande and Temane fields mainly (80 %) used by South Africa, *via* the Sasol pipeline. Massive natural gas reserves found recently in the Rovuma basin offshore in the north. Significant coal reserves	Poor internal access to electricity (25 % of population). National grid backbone to be constructed. Cahora Bassa and other hydropower sources of key importance but coal and gas also used increasingly for generation. Significant exporter of electricity, likely to increase considerably over time. Imports energy from South Africa to support an aluminium smelter
Namibia	No oil or gas reserves exploited currently, but Kudu gas field offshore (shared with South Africa) under early development. Ruacana hydropower facility operating in the North and hydropower at Baynes on	Significant importer of electricity, mainly from South Africa and historically also from Zimbabwe (Hwange thermal station). New agreement with Aggreko for importation from Mozambique

(continued)

Table 2 (continued)

Country	Energy resources	Electricity supplies
	the Cunene River is planned (shared with Angola). Coal imported to supply ageing Van Eck facility in Windhoek. Plans for 500 MW coal-fired facility at Walvis Bay	(Ressano Garcia) sourced from natural gas
South Africa	By far the most dominant country in southern Africa in terms of energy development to date. Small internal reserves of oil and gas but some imports from Mozambique and Kudu/shale gas in the Karoo may change this in the future. Heavily reliant on coal currently (global #9 in coal reserves) but poor quality leads to very high *per capita* greenhouse gas emissions. Major synthetic fuels sector. Minor hydropower, mostly already constructed	Strong coverage of the population, being enhanced over time. By far the largest electricity generator in southern Africa, about 90 % from coal (minor nuclear, hydropower and gas). Plans to increase nuclear generation to diversify the energy mix remain under review. Recent exports have decreased as domestic demand has increased and growth of generation capacity slowed
Swaziland	No oil or gas reserves but moderate coal resources. Biomass dominates the energy use in-country	Minor net importer of electricity, mostly from South Africa
Tanzania	No proven oil reserves but significant gas, some of this shared with Malawi (under dispute) and also with Mozambique in the Rovuma Basin. Likely to export natural gas in the future. Moderate coal reserves, supplemented by importation. Heavy reliance on biomass for fuel	Low population coverage (15 %). Most electricity from hydropower (60 %), with the remainder from fossil fuels. Essentially self-sufficient for electricity. Recent link to the SAPP; Tanzania represents a link between the SAPP and the Eastern Africa Power Pool
Zambia	Minor oil production and no gas. Significant hydropower, Kariba being dominant. New hydropower being planned	Net minor exporter of electricity from hydropower sources. This may change given the rapid growth in national demand
Zimbabwe	Minor oil reserves and no gas but moderate coal resources. Thermal power plants are present, with some hydropower. Marked reliance on biomass in rural areas	Net importer of electricity, mainly from South Africa. Previous Power Purchase Agreement for export to Namibia has lapsed

Table 3 Examples of major existing hydropower facilities in the northern countries of SADC

Country	Hydropower site	Installed capacity (MW)	Date of construction	Comments
DRC	Inga I and II	351 and 1424	1972 and 1982	Major refurbishment needed
DRC	Inga III	4800	2016 on?	Being developed currently
Lesotho	Lesotho Highlands Water Project	110	1998	Phase II being considered at present
Mozambique	Cahora Bassa	2075	1974	Operation delayed until 1997
Namibia	Ruacana	330	1978	Fourth turbine added in mid-2012
Zambia/Zimbabwe	Kariba	1470	1959	Two separate power stations exist
Zambia	Kafue Gorge	990	1973	The Itezhi-Tezhi Dam regulates flows

Table 4 The generation of electricity by South Africa for the year 2012/2013

Source	GWh (net)	Percentage of total
Coal-fired stations	232,749	90.5
Nuclear power station	11,954	4.6
Purchase from independent power producers	3516	1.4
Pumped storage stations	3006	1.2
Wheeling (transmission)	2948	1.1
Gas turbine stations	1904	0.7
Hydroelectric stations	1077	0.4
Wind energy	1	<0.1
Totals	257,155	100

6 Trade in Electricity and Virtual Water Transfers

It is important to note that the international trade of electrical power in southern Africa varies substantially over time. This reflects short-term changes in the capabilities of utilities to generate power, coupled to fluctuations in power demand. Some such variations are challenging to predict, e.g. unexpected problems at generating stations can reduce the electrical power available in a broader geography and can have knock-on effects elsewhere, sometimes with major consequences involving wide-scale load-shedding. This is a particular problem in circumstances where the reserve margin is low, as has been noted throughout almost all of SADC since late 2007—in large part because of the difficulties experienced in South Africa over recent years (reflecting the high proportion of SADC electricity that is generated in South Africa; see above).

These time-related variations in the international trade in electricity in southern Africa imply that any analysis of Virtual Water transfers in electrical supplies that are traded between the various countries should be considered as a general pattern, rather than in terms of absolute values which are in fact of relevance only to one particular time period. This is not especially problematic, however, as the key issue relating to Virtual Water transfers in the regional electricity network involves the very large difference in the consumptive use of water by facilities involving thermoelectric technology *versus* the larger hydropower plants.

Table 5 and Fig. 12 show an example of Virtual Water transfers in electricity traded internationally within southern Africa for the year 2012–2013, data on the trading pattern being derived from Eskom and from the SAPP Coordination Centre in Harare. Exports from South Africa to other countries occur in general through the SAPP transmission network and Eskom does not assign a specific source to any such trading, all electricity generated in-country being considered as a 'common pool'. In this circumstance, the national estimate of consumptive use of water in South Africa for the generation of electricity (in 2012/2013) was utilised to calculate the volumes of Virtual Water transferred in each trade. Relatively small volumes of Virtual Water were involved, as South Africa exhibits a low consumptive use of water in its overall electricity generation portfolio (and most of the power generated is in any event used nationally, rather than being traded). The total transfer of Virtual Water in electricity traded externally by South Africa in 2012–2013 was 19.34 MCM, as compared to 331 MCM of water used consumptively for all of the power generation in-country (5.8 % traded; the remainder used nationally).

The cross-trade in electricity between South Africa and Mozambique involves a considerable differential in Virtual Water transfers. As shown in Table 5 and Fig. 12, the 8280 GWh traded from South Africa to Mozambique in 2012–2013 involved a transfer of 11.76 MCM of Virtual Water, whilst the 6540 GWh of

Table 5 Transfers of Virtual Water in internationally traded electricity in the year 2012/2013

From	To	GWh/year (net)	L/kWh	MCM/year VW	Comments
South Africa	Mozambique	8280	1.42	11.76	No specific source identified; generic consumptive use figure employed
	Botswana	2570	1.42	3.65	
	Namibia	1780	1.42	2.53	
	Swaziland	600	1.42	0.85	
	Zambia	250	1.42	0.35	
	Lesotho	140	1.42	0.20	
Mozambique	South Africa	6540	296	1936	From Cahora Bassa

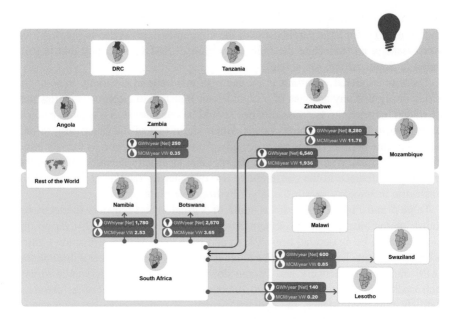

Fig. 12 The trade in electricity in 2012–2013 amongst the SADC countries

electricity derived from Cahora Bassa and traded to South Africa in the same time period implied a Virtual Water transfer of 1936 MCM.

This very large differential holds true, whether more sophisticated life-cycle analyses (or estimates of downstream effects of hydropower dams) are included, or otherwise. The distinction reflects the very great differences in the consumptive use of water by the two power generation systems, but these need also to be considered against the background of the annually renewable water resources in each country, as a whole. Thus, the degree of constraint on the water supplies is of importance in determining preferred regional patterns of electricity generation in the future.

7   Initial Reflections on the Top-Level Data

The total volume of water utilised in generating electricity in South Africa is not insignificant, at 331 MCM/year in total in 2012/2013 (2.65 % of the total annual renewable water resource nationally, making Eskom the single largest user of water in the country). However, this is altogether dwarfed by the water volumes used in the large hydropower schemes in the northern SADC countries. As noted by Beilfuss (2012), the three biggest hydropower facilities in the Zambezi River basin have fundamentally altered the total mean annual flow of the river system and have very significantly changed the seasonal flow patterns. Thus, the evaporative losses at Kariba amount to 16 % of inflows at that point, on average; those at

Itezhi-Tezhi/Kafue Gorge account for 3 % of inflows; and the evaporative loss at Cahora Bassa equates to 6 % of inflows at that point in the system. In overall terms, some 11 % of the flow of the Zambezi River system is lost by evaporation at the various hydropower facilities along its length—amounting to more than 12 km^3 of water annually, as an average. This represents by far the single largest use of water in the Zambezi River system, and rivals the blue water component of the Virtual Water volumes traded in the agricultural sector. Indeed, the Zambezi Basin IWRM Strategy recognises hydropower as the single largest water user in the Basin. (Mott-Macdonald 2008).

It is important to understand that this does not necessarily imply a preference for thermoelectric power over hydroelectric power (or *vice versa*) at any one part of the overall system within the continental SADC states. Thus, national preferences (relating to self-sufficiency/security of supply, costs, and other factors) are viewed differently by each country and the existing pattern of supply reflects this, at least to some degree. The 12 sovereign states thus face individual and collective decisions on their preferred future pattern of electricity generation and trading internationally. However, as is argued there are substantial water carbon and cost savings to be had in the conjunctive management of hydro and thermal-power.

Representatives of the SAPP have stated that electricity demand in southern Africa is increasing by more than 5 % annually (faster than the GDP growth in most of the countries), and this implies a doubling in the total demand for electricity every 14 years, amongst the continental SADC countries as a whole. In particular, very rapid increases in industrial demand for electricity between 2003 and 2007 caught ESKOM off guard. Major challenges exist in terms of satisfying the increasing demand:

- Large numbers of new generating facilities will be required. Some of these are already in construction (e.g. Inga III) or are planned (e.g. Batoka Gorge, Mphanda Nkuwa), with hydropower facilities again dominating in the northern SADC countries.[9]
- South Africa is continuing to consider its preferred energy mix, with possible increases in nuclear energy but the certainty of new coal-fired stations also being involved (the Kusile thermoelectric station having been selected to follow the finalisation of the Medupi facility), and the potential use of shale gas (although the latter remains highly controversial).
- Major new transmission systems are required, both to satisfy existing demand and to create a robust network for international trade in electricity. This is a particular focus of the SAPP, and some of the challenges that will be faced are highlighted in Chapter "Electrical Power Planning in SADC and the Role of the Southern African Power Pool".

---

[9]It is notable that when these new facilities come online, some will affect the data for consumptive water use as shown here. This will occur as new plants are constructed downstream of existing facilities which already enjoy regulated flows from upstream reservoirs, and also when some of the existing facilities are expanded in relation to their electrical output.

For South Africa, the future is especially challenging due to tightening restrictions on water availability; the inexorable increase in demand for electricity; and the current reliance of some of the neighbouring countries on exports of electricity from South Africa. The high assurance of supply that is required for electricity generation implies that the use of blue water exerts particularly severe effects on other types of water utilisation or allocation in times of drought. Inter-sectoral competition for water already exists in many of South Africa's basins (and in some of the neighbouring countries also), and the high and increasing domestic/municipal demand for water is especially notable on a country-wide basis in South Africa. In addition, the atmospheric emissions from power generating facilities in South Africa are considerable and concerns exist over greenhouse gases and effects on climate change. The possibility of the development of shale gas resources in the Karoo—estimated by the US Energy Information Administration to amount to up to 485 trillion cubic feet—remains controversial in large part due to concerns over the environmental impacts of hydraulic fracturing ('fracking').

Other countries in southern Africa exhibit varying responses to the energy sector, and the satisfaction of their own national demand (see Table 2). For example, Namibia has long been reliant on imports of electricity from South Africa, but it is predicted that this will change in the future as major new supplies are developed within Namibia (a coal-fired power station near Walvis Bay; the Kudu gas reserve in Namibian waters offshore; and the Baynes hydropower site on the Kunene River). Botswana depends heavily on its national coal resources, and also on imports of electricity from South Africa. Tanzania, having suffered load-shedding due to low water levels serving its hydropower plants (which produce almost 40 % of the country's electricity needs) is focusing on expanding thermal energy production—and the newly discovered gas supplies in the Rovuma basin offshore offer considerable future potential in this regard. Nonetheless, there are still opportunities to develop hydropower in the wetter regions of Tanzania.

Other SADC countries have limited hydrocarbon reserves and continue to rely heavily on hydropower, with major facilities planned or under development at sites such as Batoka Gorge (Zambia/ Zimbabwe), Lower Kafue Gorge (Zambia), Mphanda Nkuwa (Mozambique), and Inga III (the DRC). Extensions to existing hydropower stations are also envisaged, substantially increasing the present electricity generation portfolio of SADC as a whole.

The Grand Inga development in the DRC would fundamentally alter the pattern of electricity production in Africa, given its massive potential (variously estimated as 39,000–42,000 MW). Several types of schemes have been considered, including run-of-the-river facilities and impoundments of various forms. Although the latter would be likely to imply a higher consumptive use of water, the Congo River flow is so massive, and the use of water in the DRC is so small at the present time (the Congo's low flows tend to equate with the Zambezi's peak flows) that Virtual Water transfers should not raise problems in that instance.

A second factor that could substantially change the future pattern of electricity production in southern Africa involves the very large gas reserves discovered recently in the offshore Rovuma Basin in northern Mozambique/south-eastern

Tanzania, referred to briefly in the text above. Current estimates suggest that this will exceed 100 trillion cubic feet and the world class resource offers very significant opportunities for electricity production (with relatively low consumptive water use and atmospheric emissions). No decisions have been made as yet as to the preferred use of the gas—although the export of liquefied natural gas to the Far East appears to be almost certain for a significant proportion of the resource.

It is also possible that South Africa will proceed with the exploitation of shale gas in the Karoo. However, this option involves the use of hydraulic fracturing ('fracking'), which remains highly controversial; potentially contaminative; and relatively water intensive. No analysis of the possible use of water in fracking procedures has been made during the current work, although this could be addressed at a later stage to provide important new information of relevance to eventual policy considerations.

The impacts of climate change on water availability are difficult to estimate. The Word Bank estimated that climate change in the Zambezi River basin may reduce mean annual river flows predicted by 16 % in the upper section; 24–34 % in the middle reaches; and 13–14 % in the lower Zambezi (World Bank 2010). While most scientists now acknowledge that the precise changes to hydrological parameters that would accompany climate change are challenging to predict, any such alterations in flow patterns would be little short of catastrophic in relation to hydropower generation in the Zambezi basin.

Various alternatives are being considered in this regard, including the operation of hydropower plants as effective run-of-the-river facilities (at full reservoir levels), coordinated hydropower management (as opposed to unilateral management), and other possibilities.

Such considerations bring the importance of regional planning into the forefront of the debate. The rationale for the creation of the SAPP is instructive in this regard (Economic Consulting Associates 2009a, b). Prior to the SAPP being established, the early development of transmission systems between Botswana, Zambia and Zimbabwe was intended to reduce the reliance of the 'Frontline States' on imports of electricity from South Africa during apartheid. The dismantling of apartheid in the early 1990s brought new opportunities and in response to the 1991–92 drought which imposed severe limitations on hydropower production in the Zambezi River basin, South Africa was able to fill the gap through the existing transmission network. Other facilities brought into operation at a later time (e.g. Cahora Bassa) assisted in creating greater robustness within the regional electricity network. Nonetheless, the regional transmission network requires upgrading for the SAPP to reach its full capacity, and currently only a small portion of the demands for trades through the SAPP can be met (see Chapter "Electrical Power Planning in SADC and the Role of the Southern African Power Pool").

South Africa possessed two goals through this and the later period: initially to act as the power-house of southern Africa (and maybe even the continent as a whole); and later to be in a position to import relatively cheap hydropower from the northern SADC countries. For some eight years, a regional option of importing energy from the north at peak times, and reversing that process off peak allowed for the optimisation of hydropower and thermal power—effectively storing energy behind the hydropower dams in the Zambezi for use at peak times. However, the long term operation has been compromised, primarily due to the rapid growth in electricity demands in the region, making little available for export. The bilateral trade in electricity that pre-dated the establishment of the SAPP continues to represent the great majority of the electricity trade in the region, the initial short-term electricity market (STEM) and the more recent day-ahead market (DAM) created under the SAPP representing only very small proportions of the electricity traded internationally.

In the present context, the southern countries within SADC face a complex challenge—the need to meet sovereign demands, while also receiving benefits from regional cooperation in electricity trading (and the water-related benefits that may accompany this). Thus:

- The demand for electricity is growing more rapidly that GDP in all of the countries involved, and commonly at a faster pace than the electricity generation/transmission infrastructure can be delivered.
- There is little scope for further hydropower generation in the southern countries, with the exception of the 600 MW Baynes facility to be shared by Namibia and Angola on the Kunene River, and the expansion of hydropower production through the Lesotho Highlands scheme.[10]
- A continued reliance on coal-fired power stations in SADC as a whole will exacerbate the emissions of greenhouse gases, which are already high on a *per capita* basis, in South Africa in particular.
- Concerns raised by the Fukushima event of 2011 and by previous incidents in the nuclear power industry have cast doubt on plans for enhanced nuclear power generation in South Africa.
- Only minor scope exists for the expansion of renewable electricity generation, e.g. from wind and solar sources, and the lead times for these systems may be longer than is typical in more developed economies.

The southern countries in SADC—which experience much greater water stress than those further north—need to finalise strategic decisions on their preferred approach to electricity generation in the future. The fundamental components of this decision relate to a choice between additional thermoelectric facilities which will maintain national supplies in South Africa at least, *versus* the importation of the relatively cheaper mix of hydropower and thermally-derived electricity from the northern SADC countries (the latter, accompanied by a concomitant reduction in national energy security for the southern states). The preferred degree of regional integration represents a significant issue in this regard, and some commentators have placed this goal to the forefront of the debate (African Development Bank 2011).

---

[10]The second phase of the Lesotho Highlands Water Project was officially commenced in late March 2014. This is reported to involve a total cost of R15.5 billion, and will involve the construction of the Polihali Dam and the Kopong pumped storage supplying a further 1200 MW of hydroelectric power to Lesotho.

South Africa has therefore largely pursued its own power plan, rather than the optimised regional plan.

Despite this, the water-related implications of regional integration in energy production have received little attention in SADC, to date. However, the connections within the water, food and energy Nexus are coming into greater focus in the international arena. The data reported here on the distinctions in consumptive water use between thermoelectric and hydropower facilities in southern Africa augment this debate, and support the focus on the Nexus (as opposed to simply addressing water availability in isolation).

The key issue in relation to Virtual Water transfers in electricity involves the massive distinction between the Virtual Water content of electrical supplies derived from thermoelectric facilities and those from hydropower stations relying on large impoundments. This matter is addressed in brief in Text Box 3, which raises questions as to the preferred mix of regional electrical supplies in the future.

Notwithstanding the complexity of comparing the precise water footprints of various sources of electricity, for the purposes of the current study, certain broad inferences can be made specifically with respect to building regional climate resilience, improving regional integration, and building regional energy security:

- Given the strategic nature of electricity and its role in sustaining and building national economies, the SADC States are unlikely to significantly erode their sovereign security in terms of their national supplies of electricity.
- There is considerable scope for increased hydropower production in the Zambezi River basin, and this may be the only viable source to create greater sovereign security for electricity supplies to Zambia and Zimbabwe.
- There are substantial opportunities for increased thermal generation based on supplies from offshore gas fields in Mozambique and Tanzania.
- Hydropower production in the hotter northern States, while generally supported by higher flows and water availability, is highly vulnerable to increased temperatures and evaporation, and hence to climate change.
- Thermal power production in the generally dry south, while gradually becoming more water-efficient, remains a significant user of water especially in drought periods, and makes South Africa a substantial source of carbon emissions.
- The conjunctive and coordinated trading of electricity supplies through the SAPP can bring greater regional water/energy security and economic benefits for the SADC States, and can also reduce the carbon footprint of the region as a whole.

Text Box 3: An example concerning Virtual Water transfers in electrical supplies in Southern Africa

South Africa faces specific problems in relation to electricity generation in the future:

- The national demand for electricity continues to grow rapidly, with no reduction forecast.
- The capacity for additional hydroelectric supplies within South Africa is minor.
- Other renewable sources (wind, solar) are very unlikely to contribute significantly to South Africa's demand, especially peak demand, in the shorter term.
- Concerns exist in relation to the development of further nuclear energy.
- A continued reliance by South Africa on coal-fired facilities will increase atmospheric emissions yet further, which is not preferred. Carbon capture technologies are not yet sufficiently developed to reduce effects on climate change.
- There are concerns regarding the availability of sufficiently high grade coal and the reliance on open cast mining which affects the suitability of coal in wet weather.
- The South African government is currently exploring shale gas and fracking but this will occur in one of the drier regions of the country and remains highly controversial.
- Fresh water demand will also rise if electricity generation within South Africa is to increase in the future. The water demand of Eskom is already significant, at 331 MCM/year.

In such a scenario, a case can be made for a regional approach to electricity generation and use, with South Africa importing some of its electricity from countries to the North, most rely heavily on hydropower (and have very significant water resources). The development of Grand Inga in the DRC is one obvious option.

An alternative possibility (which is not mutually exclusive) would involve an agreement with Mozambique for electricity supplies from natural gas reserves, to be developed off northern Mozambique/southern Tanzania. However, this will increase South Africa's reliance on external electricity, and hence requires a shift in thinking from sovereign to regional security. This can, however, reduce the demand for water in South Africa.

8   Economic Accounting of Water

Economic accounts for water were developed initially as a specific component of the overall System of Environmental-Economic Accounting (SEEA), with the United Nations Statistics Division taking a leading role (SEEA 2014). The parent system was described in the handbook on integrated environmental and economic accounting, which is commonly referred to as SEEA-2003 (UN et al. 2003). The accounting system as a whole is aligned with the System of National Accounts used by most nations to generate statistics of relevance to their economies (ibid).

Economic accounts for water (the SEEAW) are one of five sub-components of the SEEA, the others addressing energy, fisheries, land and ecosystems, and agriculture. The key documents of specific relevance to water are those from 2006 and 2012, the later version having been aligned to a revision of the System of National Accounts released in 2008 (ibid).

The SEEAW includes the following water-related parameters as components of its standard format:

- stocks and flows of water resources within the environment;
- pressures of the economy on the environment in terms of water abstraction, and emissions added to wastewater and released to the environment, or removed from wastewater;
- the supply of water, and the use of water as input in a production process and by households;
- the re-use of water within the economy;
- the costs of collection, purification, distribution and treatment of water, as well as the service charges paid by the users;
- the financing of these costs, that is, who is paying for the water supply and sanitation services;
- the payments of permits for access to abstract water, or to use it as sink for the discharge of wastewater; and
- the hydraulic stock in place, as well as investments in hydraulic infrastructure during the accounting period.

Quality accounts and the economic valuation of water represent additional components of the SEEAW which remain experimental in nature at the present time, and are subject to further development.

9   Relevant Studies to Date in SADC

The SADC Economic Accounting of Water Use Project was completed in late 2010 (SADC 2010). Pilot Water Account Reports were produced for four countries (Malawi, Mauritius, Namibia and Zambia), and also for two river basins (the

Maputo and the Orange-Senqu). The standard approach embodied by the SEEAW was utilised, hence aligning the data with the System of National Accounts.

In general terms, economic accounting for water provides a conceptual framework to analyse the contribution of water to the economy of a country, and also the impact of the economy on water resources. This allows conclusions to be reached on whether water is utilised in efficient, equitable and sustainable fashions within countries (or in other distinct geographical units). Importantly, only blue water is addressed (in all its forms, including groundwater) in the present economic accounting systems, and green water, grey water and Virtual Water are not taken into account by the current techniques. However, the linkage between the water-related data and the national accounting system provides a powerful indicator of the uses of blue water in support of the economy, and the effects of the economy on blue water resources. The accounting system can be used at various levels, from individual enterprises and establishments (defined by their International Standard Industrial Classification [ISIC] denoters), to industry types; and also at the household level and higher levels of domestic use. ISIC components of specific relevance to water use within SADC include the following:

- ISIC 1-3 Agriculture, forestry and fishing;
- ISIC 5-33, Mining and quarrying;
- ISIC 41-43 Manufacturing and construction;
- ISIC 35 Electricity, gas steam and air conditioning supply;
- ISIC 36 Water collection, treatment and supply;
- ISIC 37 Sewerage; and
- ISIC 38, 39, and 45-99 Service industries.

Within these rather broad categories, economic accounting of water can be utilised to demonstrate the contribution of blue water to various components of a national or regional economy. This represents a tool of some utility in relation to blue water (and Integrated Water Resource Management) in the specific and complements considerations of the other forms of water.

Key indicators that are derived by economic accounting of water include:

- water use intensity;
- water pollution intensity;
- water productivity (relating to Blue Water only; see elsewhere in this report);
- water losses in distribution;
- the per capita water storage; and
- the average water price and supply cost.

Examples of the results of the SADC project on economic accounting of water include the following:

- All of the SADC countries allocate the majority of the blue water in use to their agricultural sector, even though this sector provides low returns to the economy by comparison to the industrial or services sectors. The agricultural sector is,

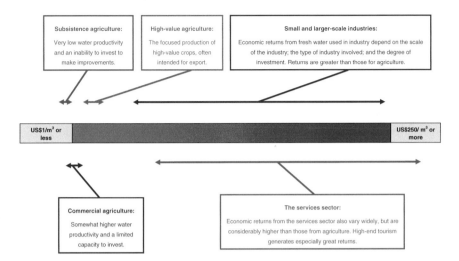

Fig. 13 Water productivity continuum, with comments on the three major sectors (Phillips 2012)

however, of particular importance for employment (e.g. 83 % of the total workforce in Malawi).

- In some countries, the sectoral comparison on returns to the economy is particularly marked. Thus, for example, a cubic metre of blue water in Mauritius provides US$0.38 as a return from the agricultural sector, but US$237 from the industrial sector. Such a pattern has been noted in other studies also (Phillips et al. 2011), as shown in Fig. 13.
- Certain countries suffer a major lost opportunity cost relating to blue water, due to poor levels of water supply and sanitation. The data for Malawi reveal that this equates to an astonishing 17 % of Gross Domestic Product (US$456 million annually), compared to a SADC-wide average of 3.2 % annually.
- Namibia performs quite well by comparison to some of the other SADC countries, in part because it has been utilising water accounting techniques since the mid-1990s (and its legislation and Vision 2030 documentation recognise the value of this approach). However, returns to the economy from the agricultural sector in Namibia are low, reflecting the highly arid nature of the country.
- Zambia is relatively water-rich, but suffers from under-development of its blue water resources (and its industrial sector, which provides high returns from blue water use). Tourism is also a notable sub-sector in Zambia, and provides very high returns for blue water allocations.
- River basin-based studies are of considerable utility in revealing the economic effects of distinct Blue Water allocation strategies. Issues relating to benefit sharing and to the appropriate pricing of Blue Water resources are also highlighted in these investigations.
- While the fact that the current techniques are restricted entirely to blue water erodes the utility of economic accounting of water approaches, the data arising

from such studies are nevertheless of importance. The SADC investigation concluded that:

- The technique can be used with pertinent results at both national and river basin level, although certain indicators are somewhat challenging to derive at the river basin level.
- A variety of indicators can be generated during the process involved in economic accounting of water use (see below), and these are of importance for decision making and policy formulation at the national and regional levels.
- The capacity to compile water accounts represents a challenge to many SADC countries, but certain states (Botswana, Namibia, and South Africa) have experience in the techniques involved and have already recognised the policy relevance of the approach.
- The compilation of economic accounts for water is best approached through the establishment of national task teams, these involving a range of governmental entities and skill bases.
- The harmonisation of certain types of data collection within SADC would assist in the compilation of economic accounts of blue water use—including a reliance on the ISIC categories in terms of industrial groupings. In some instances, hydrological data are also inadequate, and groundwater remains especially poorly characterised to date in SADC.
- Wastewater-related data are inadequate, and the water quality aspects of the approach remain under-developed as a result.
- The SEEAW approach does not address livestock as an economic unit, but blue water allocations to livestock are important in SADC.
- Economic accounting of water also fails to address the demand for blue water allocation as an 'environmental reserve' or to provide base flow in river systems.

A combination of approaches involving economic accounting of water and the techniques relating to other forms of water shows particular promise for the future. The complementary nature of the various approaches addressed in the present report is of very considerable importance, and the use of all of the forms of analysis in combination will offer a highly nuanced view of water resource management in the future. This is especially important in water-stressed systems, where allocations account for most or all of the available blue water. However, in other circumstances—exemplified especially well by Malawi within SADC, but also by Zambia and perhaps Angola—it is clear that relatively minor changes to water policy could have far-reaching implications for food security, poverty alleviation, and climate resilience.

References

African Development Bank (2011) Southern Africa Regional Integration Strategy Paper 2011–2015
Allan JA (1998) Virtual water: a strategic resource. Global solutions to regional deficits. Groundwater 36(4):545–546

Allan JA (2011) Virtual water: tackling the threat to our planet's most precious resource. London, I.B.Tauris

Beilfuss R (2012) A risky climate for Southern African hydro: assessing hydrological risks and consequences for Zambezi River Basin Dams. International Rivers

Chapagain AK, Hoekstra AY (2004) Water footprints of nations. Volume 1: main report. Volume 2: appendices. value of water research Report Series No. 16, UNESCO-IHE Institute for Water Education, the Netherlands

Economic Consulting Associates (2009a) The potential of regional power sector integration. South African power transmission and trading case study. Economic Consulting Associates Limited, London

Economic Consulting Associates (2009b) Southern Africa regional integration strategy paper 2011–2015 of the African Development Bank (2011)

FAOSTAT database available at http://www.fao.org/statistics/en/

Handbook of National Accounting on Integrated Environmental and Economic Accounting (2003) United Nations, Commission of the European Communities, International Monetary Fund, Organisation for Economic Co-operation and Development, and World Bank

Mekonnen MM, Hoekstra AY (2010a) The green, blue and grey water footprint of farm animals and animal products, value of water research Report Series No. 48, UNESCO-IHE, Delft, the Netherlands

Mekonnen MM, Hoekstra AY (2010b) The green, blue and grey water footprint of crops and derived crop products, value of water research Report Series No. 47, UNESCO-IHE, Delft, the Netherlands

Mekonnen MM, Hoekstra AY (2011) National water footprint accounts: The green, blue and grey water footprint of production and consumption. Volumes 1 and 2. value of water research Report Series No. 50, UNESCO-IHE, Delft, the Netherlands

Meldrum J, Nettles-Anderson S, Heath G, Macknick J (2013) Life cycle water use for electricity generation: a review and harmonization of literature estimates. Environ Res Lett 8(1). doi:10.1088/1748-9326/8/1/015031

Mott-Macdonald (2008) Integrated Water resources management strategy and implementation plan for the Zambezi River Basin. http://www.zambezicommission.org/index.php?option=com_content&view=category&layout=blog&id=16&Itemid=178. Accessed 18 Apr 2014

National Renewable Energy Laboratory, Colorado; and World Bank (2010) The Zambezi River Basin: a multi-sector investment opportunities analysis. The World Bank, Washington D.C.

Phillips D, Robinson and Associates (2011) The strategic focusing of potential project-based interventions in the Kagera Sub-basin of the Nile. Produced for NELSAP, 04 July 2011

Southern African Development Community (SADC) Economic Accounting of water use project. Available at http://www.sadcwateraccounting.org/1.0EconomicAccountingForWater/1.4SADCEconomicAccountingProject.aspx

South African Power Pool (SAPP) (2009) Transmission and Trading Case Study. Economic Consulting Associates Limited, London, October 2009.

Southern Africa Regional Integration Strategy Paper 2011-2015 of the African Development Bank (2011)

System of Environmental-Economic Accounting (2014) http://unstats.un.org/unsd/envaccounting/seea.asp

Torcellini P, Long N, Judkoff R (2003) Consumptive Water Use for U.S. Power Generation. Technical Report, National Renewable Energy Laboratory, Golden, Colorado

United Nations Statistics Division (2014) Accessed online on 14/03/2014 http://unstats.un.org/unsd/trade/methodology%20IMTS.htm

US Energy Information Administration (2014) http://www.eia.gov/countries/

World Bank (2010) The Zambezi River Basin: a multi-sector investment opportunities analysis. The World Bank, Washington D.C.

The Future of SADC: An Investigation into the Non-political Drivers of Change and Regional Integration

Anthony Turton

Abstract Biophysical drivers have the potential to alter the structure of inter-state relationships. Contemporary SADC is characterized by the spatial misalignment between water and economic development. The southern countries have more diversified economies, but are also water constrained; whereas the northern countries have less economic diversification, but are water abundant. The skewed nature of regional development is an artefact of history. Global climate change could disrupt these historic trading patterns, specifically when the dry portions of the region reach the hydrological limits to their internal economic growth. This disruption has the potential to reset the trade dynamics within the SADC region, specifically as each of the more economically diverse but water-constrained member states realize that food security will have to be secured at regional rather than at national level. The same is likely to occur with water and energy security, all of which can best be optimized at regional level. The question is opened about the role that Virtual Water trade can play in such a scenario. The existence of an entity known as the Southern African Hydropolitical Complex (SAHPC) becomes relevant in this regard, because conceptually it enables the current power base of each SADC member state to be analysed.

1 Introduction

The Southern African Development Community (SADC) region covers fifteen sovereign states, three of which are islands. The 12 mainland African states are linked by twenty-one river basins that cross international political borders, 15 of which are considered to be the most important in terms of socio-economic development. SADC is based on the notion of integrating national economies, along lines similar to the European Union (EU). SADC was born out of the Cold War when a

A. Turton (✉)
Centre for Environmental Management, University of Free State,
Bloemfontein, South Africa
e-mail: tony@anthonyturton.com

© Springer International Publishing Switzerland 2016   45
A. Entholzner and C. Reeve (eds.), *Building Climate Resilience through Virtual Water and Nexus Thinking in the Southern African Development Community*, Springer Water, DOI 10.1007/978-3-319-28464-4_3

number of localized wars of liberation and independence were fought across the sub-continent of Africa (Turton 2004). One of the responses to that process was the founding of the Southern African Development Coordinating Conference (SADCC) as a means of coordinating development aid in what was then known as the Front Line States (FLS) in their struggle against colonialism, capitalism and racism (Baynham 1989). The EU was also born of conflict after two wars in Europe engulfed the entire world (Nye 1971). Both of these organizations are thus based on the logic of a peace dividend that would be capable of effectively containing any localized conflict with regional integration based on trade and infrastructure as a key element. When peace broke out in the Southern African region, there was no longer the need for SADCC, so it transformed itself into SADC, with one of the first protocols being agreed by all member states focussing on the peaceful development of the many shared river basins in the region (Ramoeli 2002). The development of water resources is thus a fundamental aspect of SADC with strategic ramifications (Basson 1995; Heyns 2002; Snaddon et al. 1998; Turton 2003). This chapter will explore the non-political drivers of change in the SADC region, by focussing on water as a manifestation of the development potential and aspirations of the various member states. The linkage to energy (World Bank 1992) and food production (Africa 1984) will also be developed, within the broader context of the Water-Energy-Food-Nexus (WEF Nexus) as manifest in the concept of Virtual Water trading (Allan 2011; Earle and Turton 2003; Hoekstra and Hung 2002; Turton et al. 2000). Inherent to this is the inescapable fact that water resources are unevenly distributed within the SADC region (McDonald and Partners 1990), so in the face of global climate change (Scholes and Biggs 2004), a growing population base (Turton and Warner 2002) and the current absence of significant intra-regional trade, a series of biophysical drivers are at work. This chapter seeks to map out some of those non-political drivers by exploring possible future development scenarios that are likely to arise.

2   Assumptions

Given that this paper attempts to look into the future, there is obviously a degree of uncertainty involved. In order to meet the objective of the chapter in a meaningful way, it is necessary to base the contents on three key assumptions. These will be taken as given, merely because this set of assumptions then allows a viable set of future scenarios to be developed. The purpose is thus not to interrogate whether these three assumptions are valid or not, but rather to use them to develop a plausible argument.

The first significant assumption is that climate change is happening (Scholes and Biggs 2004) so the author is not going to argue the science. The foundation of the scenarios presented in this chapter is based on the assumption that the SADC region

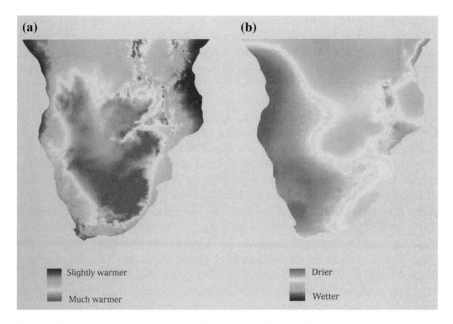

Fig. 1 SADC region and climate change (Scholes and Biggs 2004)

is likely to impact in a mosaic, best captured by the data present in Fig. 1 that shows probable future deviations from the norm in terms of temperature and precipitation. This will have significant implications on water resource availability across the region, with the potential to become a significant driver of regional integration over time. The assumption is that in essence the SADC region is likely to become hotter and drier, most notably in the hinterland.

The second assumption is that a hydropolitical complex exists (Turton 2001a, b). The implication is that there is already a set of inter-state relationships that exist, and are partially shaped by, the strategic need of each SADC member state to find security of water supply. A hydropolitical complex exists when patterns of inter-state amity (cooperation) and enmity (conflict) converge around the co-dependence on a specific shared water resource, with the overall pattern of convergence tending towards co-operation rather than conflict (Ashton and Turton 2008; Maupin 2010; Turton 2008; Turton and Ashton 2008). This is the pattern in the SADC region, so it is prudent to call the SADC region the Southern African Hydropolitical Complex (SAHPC) when referring to interstate relations over water, specifically because of the convergence around cooperation rather than conflict (Turton 2007; Turton et al. 2006b). Within the SAHPC there are two distinct classifications of riparian state and transboundary river basin. Pivotal states are those that are the most economically diversified, with water resources likely to constrain future economic growth and development. Pivotal states have a natural tendency to play a hydrohegemonic role in a given riparian relationship, by virtue of the fact that their greater diversification often translates into a political, military and economic power asymmetry in their favour. Impacted states are those that are least economically diverse, typically locked into a relationship with a pivotal state by virtue of co-dependence on a transboundary water resource, in which the power asymmetry between the various riparian states does not favour them. In the SAHPC there are four pivotal states—Botswana, Namibia, South Africa and Zimbabwe—all of which have water constrained national economies. An impacted state often has significant water resources available on which future growth can be based, but is typically lacking the investment and physical infrastructure to develop the economy (referred to as economic water scarcity in Chapter "Mechanisms to Influence Water Allocations on a Regional or National Basis"). Pivotal basins are those on which a pivotal state has a high level of dependence, and are thus regarded as a strategic issue in those countries, and which have already been fully, or almost fully, allocated. There are three pivotal basins in the SAHPC—the Incomati, Limpopo and Orange/Senqu—which can be considered as closed resources (Ashton and Turton 2008) (see Fig. 8). The other basins are known as impacted basins, because while they have water available, this cannot necessarily be developed because of the absence of investment, infrastructure and economic diversification to support a robust tax base. Groundwater resources have not yet been fully classified, but current work being done by the International Groundwater Assessment Centre (IGRAC) is showing that a large number of aquifers are in fact transboundary in nature, so their relevance is likely to be better understood in terms of the SAHPC as soon as more information is known about them. This unique pattern of distribution has a number of ramifications that are absent from the current literature. The SAHPC is presented schematically in Fig. 2 that names the water resources— both surface and groundwater—on the top horizontal axis; while the left hand vertical axis shows the riparian states. The matrix gives an indication of the hydrological foundation of the SADC region, most notably by identifying areas of convergence in the management of any specific resource.

The third assumption is that while sovereignty is known to be a challenge in any regional integration scenario, it can be managed, so it need not necessarily become a stumbling block (Turton 2002). The significance in the context of this paper is based on the core argument that global climate change will increasingly challenge state sovereignty in terms of water, energy and national food supplies, by forcing regional cooperation to the point where these three forms of security will be sourced at regional rather than national level, with the SAHPC as a key defining feature of future interstate dynamics. This might trigger a desire by SADC member states to cooperate in a way that does not erode their sovereign authority, such as via a Parallel National Action (PNA) model (Nielsson 1990; Turton 2008).

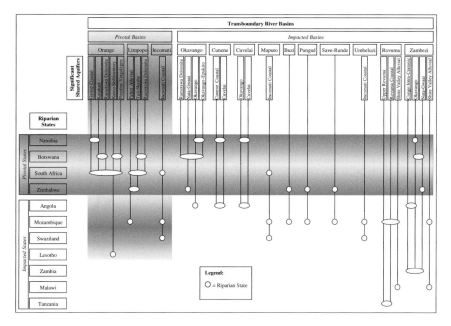

Fig. 2 Schematic representation of the relationship between significant water resources and various units of management in the SADC region structured within the SAHPC (Turton et al. 2008)

3 The Current Situation

The SADC region is characterized by a specific hydrological regime, made more complex by the fact that the majority of the area lies between the Inter-Tropical Convergence Zone (ITCZ) and the Southern Ocean, both of which drive different patterns of weather and precipitation (see Fig. 3).

This biophysical characteristic is superimposed onto a set of countries, each with different developmental trajectories, different political histories, differing legal systems that reflect previous colonial legacies and diverse natural resource endowments. Critical linkages arise from the interplay between these complex levels of scale, each driving different perceptions of national sovereign risk arising from attempts by different stakeholders (both state and non-state actors) to meet their legitimate developmental aspirations.

4 Regional Participation

The economic development potential of the SADC region is defined by the availability of water. The primary source of water is precipitation, which is highly skewed across the region. The precipitation patterns are characterized by steep

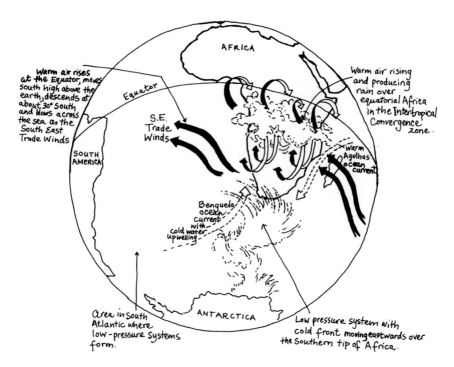

Fig. 3 Schematic representation of the two major environmental systems that drive weather in the SADC Region (Pallett et al. 1997: 13)

gradients from north to south and from east to west, with the most currently economically diverse countries being on the "wrong side" of the global average of 860 mm/year^{-1}. The data presented in Fig. 4 shows these precipitation-related facts in a dramatic way, with the red line representing the global average isohyet of 860 mm/year^{-1} and the number stated in brackets beneath each country name representing the annual average precipitation for that country.

Arising from these precipitation patterns, the SADC region has a very specific drainage system. As a result of the colonial legacy, international political borders seldom reflect hydrological management units, which in terms of 21st century thinking are the river basin, defined as the area within the physical boundary delineating the surface drainage area. This area is also erroneously referred to as the watershed, which in reality is the line demarcating the boundary of a catchment or basin, where water is 'shed' towards the drainage outlet. At continental level, Africa has 64 river basins that cross international political borders (the 63 noted by Ashton and Turton (2008) plus Lake Chilwa (see Turton et al. 2008). The 21 transboundary rivers relevant to SADC are shown in Fig. 5.

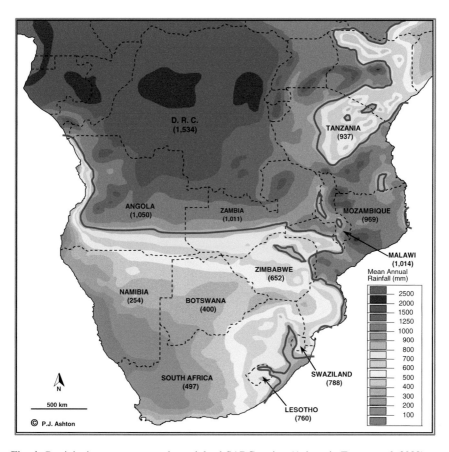

Fig. 4 Precipitation patterns over the mainland SADC region (Ashton in Turton et al. 2008)

5 Regional Hydrology

Given the highly skewed nature of the regional precipitation the SADC region also
has a very specific hydrology. This is driven by the conversion of rainfall (MAP) to
river runoff (MAR). Figure 6 shows the MAP:MAR conversion ratio for the 21
transboundary river basins in the SADC region. The horizontal axis represents
MAP with the vertical axis showing MAR. The small dots on the graph represent
individual river basins in the SADC region, with the larger dots representing
specific countries by way of comparison. It is immediately evident that while the
river basins in the SADC region differ in terms of volumetric flow, they are mostly
clustered along or below the 10th percentile (O'Keeffe et al. 1992). This means that
while the continental average MAP:MAR conversion is 20 % (UNEP 2002a), the
SADC conversion ratio is considerably less, being in most cases half of that (10 %),
often from a low precipitation base. In Virtual Water terms, this means that only

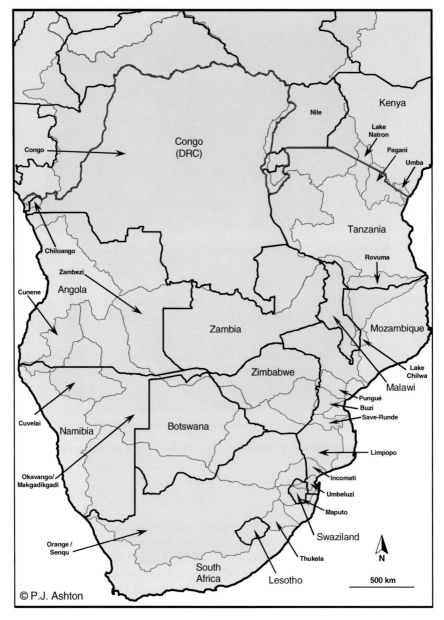

Fig. 5 Geographic distribution of the twenty-one rivers to which one or more SADC member state is riparian (Ashton in Turton et al. 2008)

some 10 % of the total water available from rainfall becomes available as blue water, with 90 % available as green water. Yet exports of agricultural goods from SADC to the rest of the world include some 17 % blue water. In some countries and

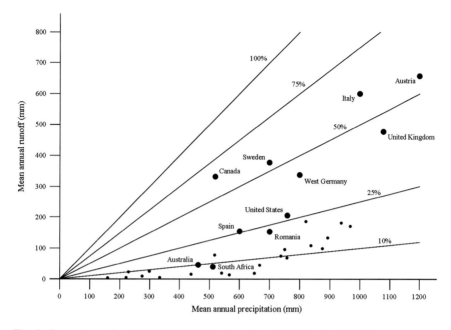

Fig. 6 Conversion ratios of MAP to MAR (redrawn from O'Keeffe et al. 1992)

products up to 70–80 % of the Virtual Water exported is blue water. It is this set of factors—a combination of climatic, hydrological, and current water use—that is a fundamental developmental constraint in the SADC region.

In the context of this chapter, it will be assumed that this MAP:MAR conversion rate will change under a future climate scenario in which the SADC region becomes both hotter and drier, reducing the proportion of rainfall that results in runoff even further. Potentially this translates into a significant driver of change within the SAHPC and is thus highly relevant to the future of regional integration.

Work done at the global scale by Shiklomanov indicates that Africa as a whole is relatively poorly endowed with surface water flowing in rivers (UNEP 2002a). The relevance of this fact becomes stark when the data presented in Fig. 4 is super-imposed onto the data presented in Fig. 6, because it indicates that the SADC region is particularly poorly endowed with water in rivers, if one factors out the impact of the Congo River. This makes the other transboundary river basins extremely important for the future economic integration of the SADC region.

While surface water is important, the significance of groundwater should not be forgotten. Groundwater in the SADC region is a vital resource, often used by rural communities as their only reliable source of drinking water. The livelihood flows derived from groundwater are thus extremely important, specifically in terms of poverty eradication, so these should not be forgotten. Table 1 lists the 22 known transboundary aquifer systems within the SADC region along with their respective riparian states (Maupin 2010; Turton et al. 2006a). The column on the right shows

Table 1 There are twenty-two known transboundary aquifer systems in the SADC region, but their geographic extent and hydrogeological characteristics are poorly described (Maupin 2010; Turton et al. 2006b)

Riparian State	Aquifer											
	Cunene Coastal	Cuvelai	Congo Coastal	Congo Intra-Cratonic	Gariep Coastal	Incomati Coastal	Kagera	Kalahari	Karoo Sedimentary	Kenya-Tanzania Coastal	Kilimanjaro	Limpopo Granulite-Gneiss Belt
Angola	X	X	X	X								
Botswana								X				X
DRC			X	X								
Lesotho									X			
Madag.												
Malawi												
Mauritius												
Moz.						X						
Namibia	X	X			X			X				
South Africa					X	X		X	X			X
Swaziland						X						
Tanzania							X			X	X	
Zambia				X								
Zimbabwe												X
States sharing	2	2	2	3	2	3	1	3	2	1	1	3

Riparian State	Aquifer										Shared aquifers
	Nata-Gwaai	Okavango	Okavango-Epukiro	Pafuri Alluvial	Pomfret-Vergelegen Dolomitic	Ramotswa Dolomite	Rovuma Coastal	Shire Valley Alluvial	Tuli-Shashe	Upper Rovuma	
Angola		X									5
Botswana	X	X	X		X	X			X		8
DRC											2

(continued)

Table 1 (continued)

Riparian State	Aquifer										
	Nata-Gwaai	Okavango	Okavango-Epukiro	Pafuri Alluvial	Pomfret-Vergelegen Dolomitic	Ramotswa Dolomite	Rovuma Coastal	Shire Valley Alluvial	Tuli-Shashe	Upper Rovuma	Shared aquifers
Lesotho											1
Madag.											0
Malawi								X			1
Mauritius											0
Moz.				X			X	X		X	5
Namibia		X	X								6
South Africa				X	X	X			X		9
Swaziland											1
Tanzania							X			X	5
Zambia	X	X									3
Zimbabwe	X			X					X		4
States sharing	3	4	2	3	2	2	2	2	3	2	

how many shared aquifers are in each country and the bottom row shows how many riparian's exist within each shared aquifer system.

If one superimposes the surface and groundwater resources available across the SADC region, then an interesting pattern of distribution occurs. Figure 7 shows a schematic representation of the distribution of water resources, both surface and groundwater, across the SAHPC. It is significant to note that the four most water constrained countries that are on the "wrong side" of the global average isohyet of 860 mm/year^{-1} (see Fig. 8)—Botswana (400 mm/year^{-1}), Namibia (254 mm/year^{-1}), South Africa (497 mm/year^{-1}) and Zimbabwe (652 mm/year^{-1}); are also the countries that share the largest number of transboundary aquifers (see Table 1)—Botswana (8), Namibia (6), South Africa (9) and Zimbabwe (4). These four countries are pivotal states in the SAHPC, and the three transboundary surface water basins that they depend on for strategic supplies of water, and which have already been fully—or almost fully—allocated (Incomati, Limpopo, Orange/Senqu), are called pivotal basins.

It becomes instructive to see where the main transboundary aquifers are. Maps to this effect are scarce and where they are found are extremely inaccurate by virtue of the paucity of knowledge about the full geographic extent of the aquifer systems. Recent work by Cobbing et al. (2008) generated a map that was further developed by Maupin (2010: 60). From this it is evident to what extent water is shared by the various SADC member states, both surface and groundwater.

With these facts having been presented, it now becomes possible to examine each of the transboundary rivers in the SADC region with a view to determining what national sovereign risk each manifests for sustaining national economic development and regional integration. The SADC region has 21 transboundary river basins to which one or more SADC Member States are riparian. These are shown in Table 2, which also indicates the existence of an inter-state agreement, the names of the respective riparian states and the classification in terms of being either perennial (a river that flows permanently) or ephemeral (a river that flows inter-mittently, mostly driven by specific episodic events giving such systems a unique hydrology and risk profile).

The 21 transboundary river basins to which a SADC member state is riparian are shown on Fig. 8. From this map it is evident that there are three broad categories of transboundary rivers. Category 1 consists of those transboundary rivers where not all of the riparian's are members of SADC. This means that the SADC Protocol is not necessarily applicable to the management of that specific river basin, but it could become the foundation for management in the future.[1] Included in this category are the: Chiloango, Congo, Lake Natron, Nile, Pagani and Umba basins. Category 2 consists of those transboundary rivers where all riparian's are members of SADC, so the management of those systems is subject to the SADC Protocol. This consists of two distinct subsets. Category 2a consists of rivers that have

---

[1]Some international customary law principles contained in the protocol could nevertheless be applied even under the present context.

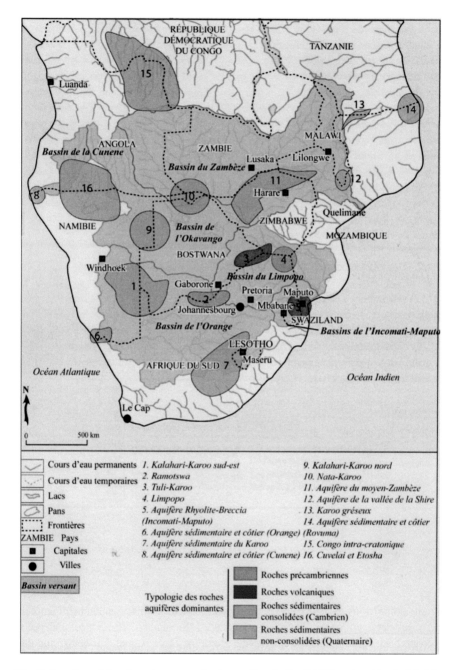

Fig. 7 Location of transboundary groundwater aquifers in the SADC region (As depicted by Maupin (2010: 60) using Turton et al. (2006b), Cobbing et al. (2008) and IGRAC (2009) as primary sources)

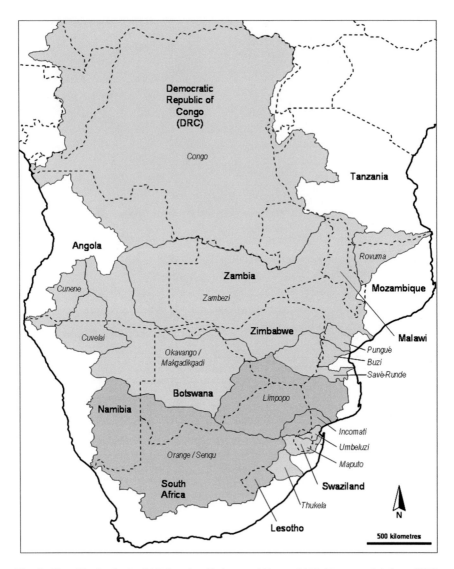

Fig. 8 Closed basins in the SADC region (Ashton and Turton 2008; Turton and Ashton 2008)

significant portions of their basin in each riparian state so joint management is vital; this sub-set consists of the: Cunene, Incomati, Limpopo, Maputo, Okavango/Makgadikgadi, Orange/Senqu, Rovuma, Savé-Runde, Umbeluzi and the Zambezi basins. Category 2b consists of rivers that are fully within SADC territory and thus under the jurisdiction of the SADC Protocol, but are characterized by basins where the largest proportion of the resource lies in one country. As a result, joint management is not critical and might even be impractical. This sub-set includes the: Buzi, Pungué and the Thukela basins. Category 3 consists of rivers

Table 2 Transboundary River Basins to which one or more SADC member state is a Riparian (Turton et al. 2008)

Transboundary river basins to which one or more SADC member state is a riparian			
Basin name	Agreement	Type	Riparian states
Buzi	No	Perennial	Mozambique, Zimbabwe
Chiloango	No	Perennial	Angola, DRC
Congo	Yes	Perennial	Angola, Tanzania, Zambia
Cunene	Yes	Perennial	Angola, Namibia
Cuvelai	Yes–no RBO	Endorheic and Ephemeral	Angola, Namibia
Incomati	Yes	Perennial	Mozambique, South Africa, Swaziland
Lake Chilwa	No	Endorheic	Malawi, Mozambique
Lake Natron	No	Endorheic	Kenya, Tanzania
Limpopo	Yes	Perennial	Botswana, Mozambique, South Africa, Zimbabwe
Maputo	Yes	Perennial	Mozambique, South Africa, Swaziland
Nile	Yes	Perennial	DRC, Kenya, Tanzania
Okavango/Makgadikgadi	Yes	Endorheic	Angola, Botswana, Namibia
Orange/Senqu	Yes	Perennial	Botswana, Lesotho, Namibia, South Africa
Pagani	No	Perennial	Kenya, Tanzania
Pungué	No	Perennial	Mozambique, Zimbabwe
Rovuma	Yes–no RBO	Perennial	Mozambique, Tanzania
Savé/Runde	No	Perennial	Mozambique, Zimbabwe
Thukela	No	Perennial	Lesotho, South Africa
Umba	No	Perennial	Kenya, Tanzania
Umbeluzi	Yes	Perennial	Mozambique, South Africa, Swaziland
Zambezi	Yes	Perennial	Angola, Botswana, Malawi, Mozambique, Namibia, Zambia, Zimbabwe

that have specific hydrological regimes, which are not conducive to the construction of large dams, mostly being endorheic in nature, but sometimes also ephemeral typically being linked closely to groundwater systems. This means that a disproportionately large number of livelihoods are dependent on a water supply that is often highly irregular and erratic, and where the management of these systems is very complex, often linked to endemic poverty, and mostly under-funded. Included in this category are the: Cuvelai and Lake Chilwa basins. Of these transboundary river basins, a number are closed out, meaning that more rights have been allocated to these systems than actual water available at a high assurance of supply. Basin

Table 3 Physical description of the major transboundary rivers in the SADC region

Basin name	Total basin area (km²)			River length (km)	Mean annual runoff (Mm³/year⁻¹)
	Pallett	UNEP	Wolf		
Buzi	31,000	–	–	250	2500
Congo/Zaire	3,800,000	3,669,100	3,669,100	4700	1,260,000
Cunene	106,500	110,000	–	1050	5500
Cuvelai	100,000	–	–	430	Ephemeral
Incomati	50,000	46,000	46,000	480	3500
Limpopo	415,000	414,800	414,800	1750	5500
Maputo	32,000	30,700	30,700	380	2500
Nile	2,800,000	3,038,100	3,038,100	6700	86,000
Okavango/Makgadikgadi	570,000	706,900	706,900	1100	11,000
Orange/Senqu	850,000	945,500	945,500	2300	11,500
Pungué	32,500	–	–	300	3000
Rovuma	155,500	151,700	151,700	800	15,000
Savé-Runde	92,500	–	–	740	7000
Umbeluzi	5500	10,900	10,900	200	600
Zambezi	1,400,000	1,385,300	1,385,300	2650	94,000

Data source Columns 2, 5 and 6—Pallett et al. (1997); Column 3—UNEP (2002b); Column 4—Wolf (2006)

closure thus poses a unique set of sovereign risks, so it is important that these specific systems be distinguished from the rest.

The hydrological data for the major transboundary river basins in the SADC area is shown in Table 3. In this regard it must be noted that most sovereign states consider hydrological data to be sensitive and it is thus classified in many cases, which means that accurate data is not in the public domain. The data in Table 3 is thus the best available public domain data, used for illustrative purposes only in the context of this paper.

6 Hydraulic Infrastructure Development

Cognizant of the fact that the logic underpinning this paper suggests that water is one of the foundations of the economic development potential of the state, it becomes instructive to assess the level of hydraulic infrastructure in the region. Figure 9 shows the number and location of major dams across the SADC region.

From the data shown on Fig. 9 it is immediately obvious that there is a stark misdistribution of hydraulic infrastructure in the SADC region, with a dense clustering of dams in two of the pivotal states of the SAHPC—South Africa and Zimbabwe. When this data is superimposed onto the precipitation data shown on Fig. 4, then it is obvious that these economies are capturing the resource from the

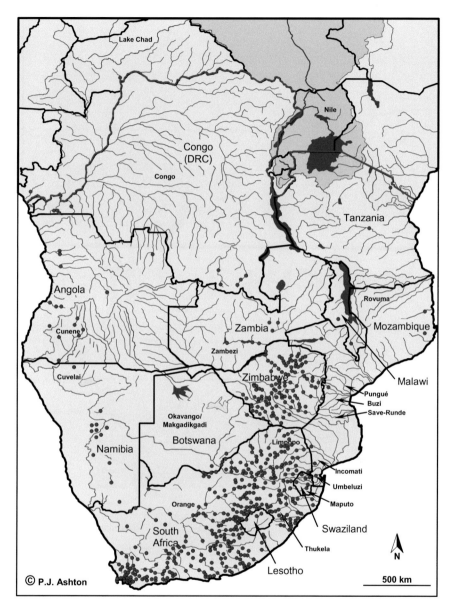

Fig. 9 Distribution of large dams in the SADC region (Ashton in Turton et al. 2008)

better-watered eastern portion of the subcontinent (Homer-Dixon 1994, 1996). The notion of resource capture by the pivotal states is even more apparent when one examines the location of current and future inter-basin transfer (IBT) infrastructure shown on Fig. 10 (Snaddon et al. 1998).

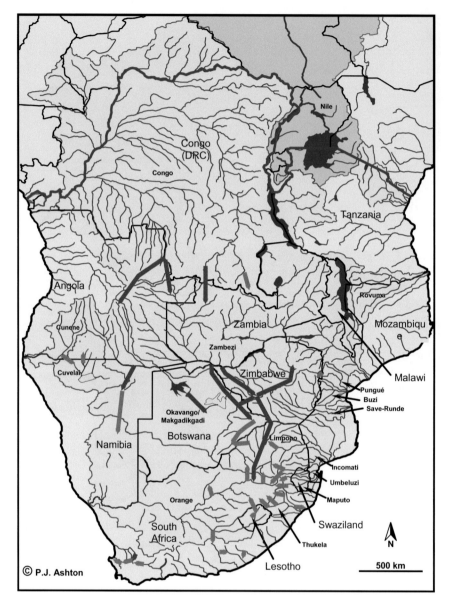

Fig. 10 Distribution of current (*orange*) and future (*purple*) IBT's in the SADC region (Ashton in Turton et al. 2008)

The data shown on Fig. 10 demonstrates a harsh reality in the SADC region, namely that while the economic development is in the dry South, the water in the wetter North will increasingly be sought as hydrological constraints to development become apparent. This is a critical piece of evidence in the logic of the SAHPC, because in essence the four pivotal states will increasingly be confronted by a sovereign security choice: either create water, energy and food security at national level by capturing the (blue) water resources of the north; or seek water, energy and food security at regional level by means of a shared vision of future infrastructural development based on Virtual Water trade using both blue and green water resources.

This is the essence of the biophysical facts that will increasingly become evident as climate change introduces a new level of risk that can only be managed by means of cooperation within the SAHPC via the institutions of SADC.

7 Broader Socio-economic Factors

Given the hydrological dilemma within the SADC region noted above, it is now necessary to analyse some of the socio-economic factors that are currently at play. Figure 11 shows the relationship between GDP per capita expressed as a function of the percentage of water imported.

Three of the pivotal states in the SAHPC are at the upper range of GDP per capita in the region, and are generally above the trend line for the scatter plot.[2] Significantly the impacted states that are also generally water abundant (Mozambique, Zambia and Angola) are lower down the GDP scale, with a significant spread between them. Angola is significant in this regard, because the annual average precipitation is 1050 mm/year^{-1} (Turton et al. 2008), making it the second wettest country in the mainland SADC region, beaten only by the DRC (with 1534 mm/year^{-1}). More importantly the GDP per capita in that country has the potential to grow in terms of the future scenario suggested in this paper, raising the question of whether Angola would choose to import its food, or grow its own. Mozambique is also important because it is a relatively water-abundant country, located on the eastern seaboard and thus in intimate contact with the warm Mozambique current. The annual average precipitation is 969 mm (Turton et al. 2008), making it the fifth wettest country in the mainland SADC region. It is riparian to the largest number of transboundary rivers in the SADC region, but it is a classic downstreamer in terms of geographic location, with a long history of being dominated by the pivotal states upstream (South Africa and Zimbabwe). It also has the potential to benefit greatly from regional integration in terms of the future scenario suggested in this paper, potentially to the point where it can break out of its relatively weaker riparian position. Zambia is also significant because the annual average precipitation is 1011 mm/year^{-1}, on a par with Angola and only slightly less than the DRC. The southern portion of Zambia where most of the agricultural

---

[2]While the trend line will be heavily influenced by the few outliers there is a general relationship between GDP and Virtual Water imported, with the exceptions being richer territorially large states (US, Canada and Australia) which are below, rich small states (UK, Mauritius) which are above and the desert countries which cannot grow their own food.

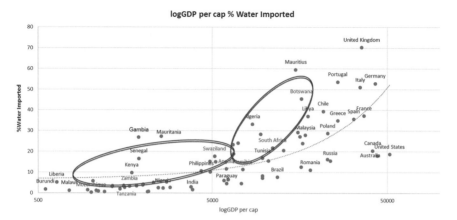

Fig. 11 Relationship between GDP per capita expressed as a function of the percentage of water imported (CRIDF dataset)

productivity lies is drought-prone however, and it is here that internal problems lie for the country—making it relatively blue water dependent. When analysed in terms of global climate change scenarios, it is evident that northern Zambia is likely to become hotter and wetter by 2050 (Scholes and Biggs 2004), which is somewhat of a better situation than most of South Africa (Scholes 1998).

When the regional data is shown as a relationship between GDP per capita and wellness (Fig. 12), then a more nuanced dimension emerges.

Three of the pivotal states in the SAHPC (Botswana, Namibia and South Africa) are once again on the upper range for the region, but are all below the global trend line for wellness. This is due to lower life expectancy due to HIV, but also expectations due to a high Gini. More significantly, three of the most water abundant of the impacted states (Angola, Zambia and Mozambique) are low on the GDP scale for the region, but straddle the global trend line on the wellness scale, suggesting that life expectancy and satisfaction with life is higher than expected for the GDP.

The Social Progress Index (SPI) measures the extent to which countries provide for the social and environmental needs of their citizens. 52 indicators in the areas of basic human needs, foundations of wellbeing, and opportunity show the relative performance of nations. This dataset has a clustering of three pivotal states in the SAHPC (Botswana, Namibia and South Africa) on the upper end of the regional range of the SPI, but all below the global trend line, again perhaps due to decreased lifespan due to HIV and the higher expectations due to high inequality. Conversely, the three most water abundant of the Impacted States in the SAHPC are low down in the regional range for the SPI, but above the global trend line, possibly because countries with low rainfall may mean more people need to migrate to where the rainfall is or where the cities are, where densely populated areas increase the risk of

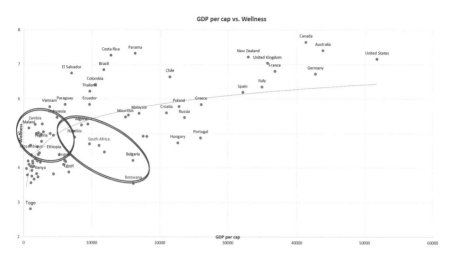

Fig. 12 Relationship between wellness and GDP per capita (CRIDF dataset)

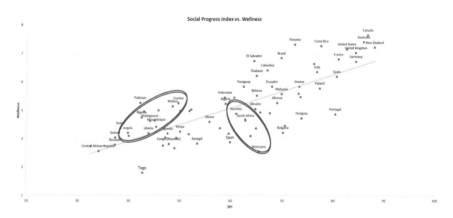

Fig. 13 Relationship between wellness and social progress (CRIDF dataset)

communicable disease. This is significant in terms of the future scenario presented in this paper.

An interesting picture emerges when one compares GDP per capita with the Human Development Index (HDI) non-income component as shown in Fig. 14. The HDI is a composite statistic of life expectancy, education, and income indices used to rank countries into four tiers of human development.

Yet again the three pivotal states in the SAHPC are clustered along a similar non-income HDI scare, but more importantly, they are significantly lower than the global trend line, again due to higher HIV rates. This suggests a large potential dividend needs to be unlocked to bring them into the global trend. The rest of the SADC countries are low on both the GDP scale and the non-income HDI scale.

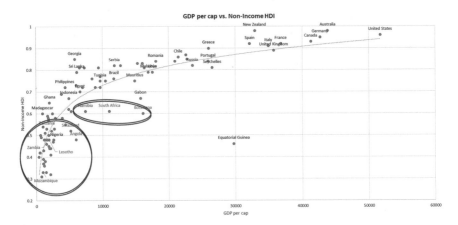

Fig. 14 Relationship between GDP per capita and the non-income aspects of the human development index (CRIDF dataset)

The regional data is presented as a function of both the Happy Planet Index (HPI) and the GDP per capita in Fig. 15. The HPI uses global data + B12 on life expectancy, experienced well-being and Ecological Footprint to calculate this. The index is an efficiency measure, ranking countries on how many long and happy lives they produce per unit of environmental input.

The stark reality of the SADC region is manifest as all of the member states score below the median in terms of the HPI. Two of the pivotal states (Botswana and South Africa) score above the median for GDP per capita, but are also the lowest in terms of HPI. Namibia, the third pivotal state, is on the threshold of the

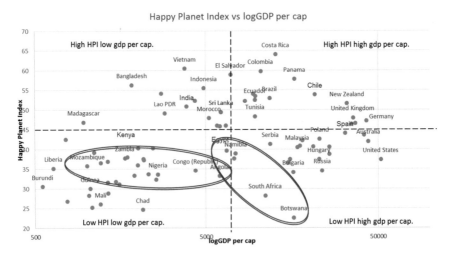

Fig. 15 Matrix showing the relationship between GDP per capita and the happy planet index (CRIDF dataset)

GDP per capita median, scoring high on the HPI scale, which would suggest that these countries have a higher impact on the planet that would be suggested by their GDP. The three most water abundant SADC countries in the dataset all occur in the quadrant with low GDP per capita and a low HPI.

The renewable internal freshwater resources per capita (cubic meters) refers to internal renewable resources (internal river flows and groundwater from rainfall) in the country. Renewable internal freshwater resources per capita are calculated using the World Bank's population estimates, while the water footprint is defined as the total blue, green, and grey amount of water consumed per person *per annum* (Mekonnen and Hoekstra 2011). The three most water-abundant SADC members in this dataset (Angola, Mozambique and Zambia) are all on the upper scale for the region. Namibia presents as an anomaly, because the reason it scores in the upper range is based on the fact that significant volumes of water flow in the rivers that form the northern and southern borders of the country, but these are not necessarily capable of being utilized or developed for a variety of complex reasons. The low population base in Namibia also skews this data significantly. None of the SADC countries score very highly on the water footprint scale when compared globally. Those that do are generally exporters of Virtual Water embedded in crops such as sugarcane (Swaziland). Malawi is an interesting case in this context, because it has a high water footprint, while also scoring low in terms of renewable water per capita. Both Swaziland and Malawi stand out in terms of the blue water component of their exports, and are relevant to the scenario presented below.

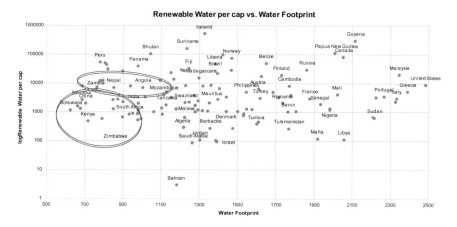

Fig. 16 Data showing renewable water per capita expressed as a function of the water footprint of the state (CRIDF dataset)

8 Potential Future Scenarios

Given the known current situation presented above, in the context of the three assumptions—that climate change is likely to make the SADC region both hotter and drier; that a hydropolitical complex exists in the region that has already structured inter-state relations in the context of water resource management; and that sovereignty need not necessarily be a stumbling block because models that promote cooperation without eroding state independence do exist—we can start to construct a set of scenarios to guide our forward thinking.

The Inter-Tropical Convergence Zone (ITCZ) funnels warm moist air across the SADC region in the summer months (see Fig. 17). It is the existence of this ITCZ weather pattern that accounts for much of the rain in the South African Highveld in summer, because it is channelled down south over the Kalahari Desert when an upper atmosphere low pressure trough forms. It is the formation of this upper atmosphere trough that makes South Africa vulnerable to climate change and increased variability, because a small disturbance in this process can alter rainfall patterns over much of the hinterland of the mainland SADC region. What happens in Angola and northern Zambia is thus of major importance to the economic activity of the SADC region, because rainfall originating from that country sustains the national economies of many countries. In terms of our future scenarios for the SADC, we will therefore assume that climate change will create a disturbance to this specific weather system that will reduce the amount of precipitation entering the atmosphere above the region. While we cannot predict with certainty, this is also likely to reduce the conversion of rainfall to runoff, effectively forcing the MAP: MAR ratio in the larger rivers to converge around the 10th percentile shown in Fig. 6 while also increasing the variability.

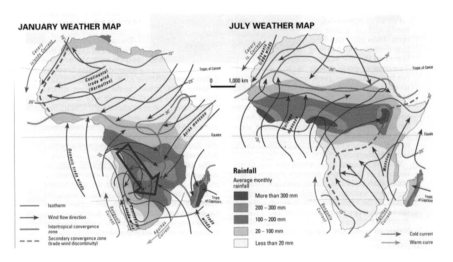

Fig. 17 Inter-tropical convergence zone (ADB 2000: 23)

The impact of this will be dramatic, because the 860 mm/year^{-1} isohyet shown on Fig. 4, will be forced to shift northwards across Angola and Zambia, and eastwards across southern Mozambique and Zimbabwe. This will change the precipitation pattern as follows:

- The rivers on which two Pivotal States in the SAHPC are totally dependent on (Namibia and Botswana) all rise on the Bié Plateau and include the Cunene, Cuvelai and the Kavango/Okavango (Fig. 5). This triggers economic stagnation as the assurance of supply is reduced and localized water shortages start to accompany electricity blackouts. Unemployment grows and social instability starts to take hold. This is particularly evident in northern Namibia where the loss of the Cuvelai impacts a large percentage of the total Namibian population. Major tributaries of the Zambezi are also affected, notably the Luiana, Cuando, Lunge-Bungo and Luena on which Zimbabwe (another pivotal state) is highly dependent. Of great strategic significance however, is the fact that an IBT from the Congo River, discharging water on the Bié Plateau (see Fig. 10), can thus be linked to increased water security for all of these rivers, with major benefits accruing to Namibia, Botswana, Zimbabwe and even Mozambique and potentially South Africa. The Proposed IBT's from the Cassai to the Cuito/Okavango, Cassai to the Zambezi and the Lualaba to the Zambezi all show the significance of the Bié Plateau in Angola (Basson 1995; Heyns 2002; Snaddon et al. 1998). This aspect, when combined with the dormant hegemonic status of Angola, is a significant factor in understanding how the hydropolitical dynamics of SADC will change, with Angola potentially becoming the dormant hydropolitical superpower in the SADC region.
- The eastwards shift of the 860 mm/year^{-1} isohyet will start to increase ambient air temperatures in the north and east of South Africa (a pivotal state in the SAHPC). This increases evaporative losses in Limpopo and Mpumalanga, both centres of energy production now and into the mid-term future. This means that South African energy security is placed under increased pressure. Swaziland is placed under increased pressure by South Africa, whose demands upstream leave less water for use by downstream riparian states. Significantly however, Mozambique also experiences an increase in demand for water, most notably from the capital city of Maputo, whose population growth and economic development demand more water than is available at a high assurance of supply level. This means that Swaziland is squeezed by the upstream riparian and regional hegemon—South Africa—and downstream riparian with a growing demand—Mozambique. The heavy dependence of Swaziland's economy on blue water (irrigated sugar) exacerbates this stress. This leaves South Africa increasingly vulnerable, but more importantly, increasingly at risk of meeting its energy needs from Mpumalanga and Limpopo provinces without re-allocating water from the irrigation sector. This means that the Greater Soutpansberg coal mining projects, currently in the exploration phase, will be unable to be realized simply by virtue of the fact that there is insufficient water available to sustain the mining process. In turn this means that nuclear power becomes an increasingly

attractive option, but opposition to this grows at the very same time that national food security declines and unemployment increases. This changes South Africa's position in the Southern African Power Pool (SAPP), as it increasingly recognises the value of hydropower in the north. Zimbabwe, another pivotal state, is also affected, because the climatic conditions that give rise to rainfall are similar. This means that the already semi-arid southern portion of Zimbabwe becomes even more arid, with the Kalahari Desert encroaching from the west but reaching deeper into Zimbabwe than at present. Zimbabwe's sugar industry and biofuels in the Save basin is put under increasing stress, affecting flows into Mozambique. The small pocket of well-watered land around the Mazoe Valley shrinks, with farm production in this specific area of Zimbabwe declining. The firm energy outputs of hydropower projects such as the Batoka Gorge in Zimbabwe, and the Mepando Uncua in Mozambique, need to be reviewed in the context of lower system yields, altering the energy planning in these countries significantly (see Fig. 18). This combines with the collapse of food production in South Africa and Mozambique, giving rise to a localized form of out-migration driven by loss of livelihood in the subsistence farming sector (Homer-Dixon 1994, 1996; Percival and Homer-Dixon 2001). Commercial farms, already under pressure from a variety of issues such as indigenization and land restitution, also lose production. The culmination of these two factors is a loss of national food security, which given South Africa's dominance in food exports to the rest of SADC, is cascaded upwards to the SADC region, which if left unmanaged, could start to resemble the type of situation currently occurring in the Horn of Africa centred on Somaliland in Kenya.

- The shifting of the 860 mm/year^{-1} isohyet has a profound impact on trans-boundary aquifer systems (see Fig. 14) as recharge is significantly reduced (Cavé et al. 2003). This occurs at the same time as these systems are being developed for poverty eradication purposes, so this translates into a localized series of humanitarian crises. The most significant aquifer system to be negatively impacted is the Cuvelai/Etosha (No. 16 on Fig. 7), which is one of the densest areas of human settlement in Namibia. This becomes the foretaste of things to come as donor agencies become aware of the problem. Other aquifers

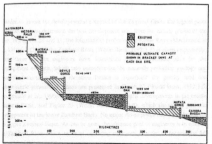
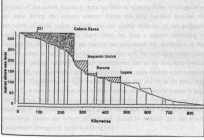

Fig. 18 Profile of the middle (*left*) and lower (*right*) portions of the Zambezi River basin indicating existing and potential dam sites (World Bank 1992: 3-3-5)

follow suit, with the Nata/Karoo and Zambezi Aquifers (No's. 10 and 11 respectively on Fig. 7) the next to manifest as a problem. Other aquifers are also affected, with the Tuli/Karoo and Limpopo Aquifers (No's 3 and 4 on Fig. 7) affecting planned mining operations and wildlife tourism destinations in a way that reduces employment for local people, further marginalizing these already vulnerable communities. The Rhyolite Breccia Aquifer (No. 5 on Fig. 7) is very significant because the peri-urban population of Maputo is directly impacted. This aquifer becomes saline as it is overdrawn, allowing salt water to intrude at the very same time as the city of Maputo faces a water crisis from the reduced flow of the Incomati, Maputo and Umbeluzi River basins (TPTC 2008) (see Fig. 5). This places considerable pressure on Swaziland, already in crisis because of increased evaporative losses and eutrophication arising from microcystin contamination of the dams. South Africa is also placed under pressure in the same systems, given their existing reliance on the Incomati and Maputo Rivers for their energy supply in Mpumalanga.

- The increase in ambient air temperature across the SADC region has a major impact on the dams located mostly in South Africa and Zimbabwe (both pivotal states in the SAHPC) (see Fig. 9). Evaporative losses increase significantly, lowering system yield, but the increase in the temperature of the water in these dams drives an increase in eutrophication (Oberholster and Ashton 2008). This will cause an increase in cyanobacterial blooms potentially resulting in micro-cystin toxin being released into the food chain, posing particular risk for vul-nerable communities with compromised immune systems (Abe et al. 1996; Bradshaw et al. 2003; Codd et al. 1999; Doyle 1991; Falconer 1998, 2005; Humpage et al. 2000; Oberholster et al. 2004, 2008; Ueno et al. 1996). This triggers a series of health-related crises in the four pivotal states, all of which are highly dependent on dams for strategic storage of water. Irrigation farming is placed under pressure, most notably because of the improved research into microcystin deposition onto crops, but also because increased evaporative losses cause salinization of the soils over time, reducing crop yield significantly. Swaziland is also affected, given the high reliance on irrigated agriculture for that country, and GDP is directly affected as exports diminish on the back of media coverage about microcystin contamination of produce.

As the regional hydrology changes, so too does the climate, with the Kalahari and Karoo generally expanding northwards as far as southern Zambia and east-wards as far as Mozambique. This has a direct impact on national food production in the four pivotal states of the SAHPC. Initially seen as a crisis, realization starts to dawn on decision-makers that Virtual Water trade is a regional solution, so food security is increasingly found at regional rather than national level in these four countries. Regional food production moves to the following countries:

- The DRC is well watered and has vast landscapes that are capable of being tilled. International interest is triggered as multinational agribusiness corpora-tions secure land in order to seize market share for the future. A parallel process occurs with smallholder agriculture being organized into cooperative movements capable of moving produce to the markets of the south before spoiling. South African and Zimbabwean commercial farmers, displaced because of land restitution issues in those two countries, migrate northwards and set up successful operations. The challenge for the DRC is the lack of infrastructure and the relative weakness of the central government, so security remains a constraint and development of the agricultural resource-base is initially slow and patchy.

- Angola is also well watered, so it starts to emerge as a new regional hegemon by virtue of the fact that it controls the Bié Plateau, and with that the headwaters of the Cunene, Cuvelai, Kavango and parts of the Zambezi. IBT's from the Congo into these systems enhances the prestige and power of the country. However, the Angolan economy is generally outward looking, based mostly on mining and energy, with trading partners outside of the SADC region. This inhibits Angola to reach its full potential as a hegemonic state in the SAHPC, but it does not curtail the natural processes at work at regional level. Large commercial farms start to develop in the central and northern portion of Angola as the desert encroaches from the south-west. Small-scale farming co-exists with large commercial farms similar to the DRC. The outward looking nature of the Angolan economy is reflected in the relative absence of transport infrastructure, so while the agricultural potential is good, connectivity to the markets of the south remains a challenge.

- Northern Zambia is also well-watered, with some indications that this might even improve under climate change conditions (Scholes and Biggs 2004). While the south sees some encroachment of the Kalahari, the rest of the country starts to boom as the agricultural potential is realized, and north—south transport infrastructure is established. The soils are good and the population density is relatively low, giving rise to a rapid expansion of large commercial farming operations. The logistical infrastructure, while not extensive, does exist to the extent that upgrades can rapidly connect the production centres in Zambia with the markets in the south. This transportation corridor is therefore prioritized with the construction of a bridge across the Zambezi at Kasane into Botswana to avoid the bottleneck at Beitbridge. Zambia therefore becomes one of the first movers in the climate change scenario, because all of the primary drivers are already in place—good soil, abundant water, existing transport infrastructure along the north-south axis. Consequently Zambia starts to emerge as a regional economic power as food processing corporations establish the first significant industrial diversification outside of mining.

- Mozambique also becomes a significant country under the scenario of climate change. While the south becomes more desertified as the Kalahari encroaches eastwards, the northern Rovuma Basin is well watered with good soils. The recent discovery of coal and gas in this area creates an impetus for the development of infrastructure, but this is initially focussed on east-west connections leaving the existing north-south road as the only significant transport corridor to the food markets. Revenues to the fiscus from mining and gas are wisely invested into north-south road, rail and electricity transmission infrastructure, so Mozambique starts to compete with Zambia as a significant food producing area, exporting both to the increasingly water stressed southern SADC countries and world-wide.

The shift in the epicentre of regional food production marginalizes the following countries:

- Malawi, with a heavy reliance on blue water in sugar is unable to translate its relatively favourable geographic position into a regional comparative advantage because of the relatively small size of the country along with the very high hydraulic density of population. Malawi therefore becomes a food importer over time with increasing levels of poverty potentially driving a population migration across the SADC region.
- Swaziland sees a significant change in its regional comparative advantage in this scenario. Climate change causes reduced flows in the rivers that sustain irrigated agriculture, with an increase in the grey water content of its sugar export, and making its exports less attractive on global markets. The reduction in precipitation also plays a role. Swaziland starts to resemble Malawi with population pressure and job losses in the agricultural sector causing an increase in poverty. As with Malawi, there is out migration pressure over time.
- Lesotho, already under pressure, becomes increasingly marginal in economic terms as the mines in South Africa shed jobs. Precipitation in the highlands is affected by the eastward shift of the 860 mm/year^{-1} isohyet, but the orographic nature of the rainfall means that system yields are not too badly reduced. Lesotho joins Malawi and Swaziland as a country in trouble as donor aid is ramped up over time. Lesotho reaches out to its ally Botswana, connected culturally and linguistically, as a water transfer is planned for Gaborone to provide an alternative revenue stream for the government.

The shift in epicentre of regional food production is likely to be felt by the four pivotal states in the SAHPC as follows:

- South Africa retains its status as the largest economy in the region, although the intra-SADC differences are reduced. Nonetheless, South Africa remains a significant industrial power. However, the loss of national food security is countered by an increase in the trade of Virtual Water in food products produced in the north. The existing road and railway infrastructure connecting South Africa to Mozambique and Zambia is rapidly upgraded, stimulating the manufacturing sector to produce goods for sale up north. Border crossings are streamlined in order to facilitate the rapid and unhindered flow of produce down south. Infrastructure investment gives Mozambique and Zambia an initial advantage as regional food producers. However, benefit is leveraged by both Zimbabwe and Botswana for the same reason. Zimbabwe is significant by virtue of the existing road and railway line northwards to Zambia. Botswana, on the other hand, is less well connected, but it leverages the growing South African dependence on food production in the north to develop its own infrastructure, most notably by upgrading the road to, and bridge across the Zambezi River at Kasane. South

Africa achieves some form of redundancy in its dependence on the goodwill of neighbouring states. As this infrastructure develops, so too does the northward trade of goods manufactured in South Africa. This drives an increase in intra-SADC trade, creating opportunities elsewhere in the region.

- Zimbabwe is increasingly vulnerable for a variety of reasons, but it survives by leveraging its gatekeeping position on the one major north-south transport infrastructure between South Africa and Zambia. The loss of food production in the country is not easily translated into Virtual Water trade because of the absence of foreign currency arising from reduced industrialization. Given the complexity of this issue the country lurches forward, but slowly starts to lose its hegemonic status as infrastructural development takes place around it rather than through it.
- Botswana is initially highly vulnerable, but strategically positioned to benefit from the changing regional dynamics. On the one hand its benefits by developing transport infrastructure bypassing Zimbabwe. It creates an inland port at Kasane by establishing a bonded warehouse facility. Early infrastructure investment starts to attract traders who increasingly set up businesses in the Kasane area. As the bridge across the Zambezi is developed, a high speed border crossing is introduced to attract logistics companies away from the Zimbabwe route, reducing the number of border crossings needed. Botswana also builds redundancy into its water supply by developing the North-South Carrier to the point where it can connect into the Zambezi at Katombora Rapids (see Figs. 10 and 18). Yet again Zimbabwe is bypassed, giving Botswana control over water delivery to Bulawayo and potentially Pretoria. Botswana also leverages its advantage on the various river basin organizations it is active in, to foster a cooperative approach to solution seeking. In this role it can broker power between various States, and given its existing dependency on imported Virtual Water, it more readily adopts the concepts into water and development policies. The linkage between Botswana and Lesotho is expanded as the possibility of transferring water from the Highlands to Gaborone is taken through a series of feasibility studies.
- Namibia is highly vulnerable because the majority of its water resources are located on either the northern or southern borders. While it has a low population base, the heavy population settlements in the north are severely impacted as the 860 mm/year^{-1} isohyet moves northwards. The first crisis occurs when the Cuvelai basin dries up for an extended period, placing additional pressure on the IBT from the Kunene back into Ongiva in Angola. However the yield from the Kunene is also affected, placing pressure on this alternative water supply scheme while also reducing the viability of the planned hydro-electric dams downstream of Calueque. The loss of food security is not readily translated into an expanding industrial base, so Namibia is increasingly marginalized in this scenario because of the low tax base to fund infrastructural development.

On the energy side of the equation, this shift in the food production epicentre of the SADC region has profound implications. Growing differences in firm versus average power from the Zambezi, and growing local demands in Zambia also shift the playing field. Broadly speaking these drivers play out as follows in the four pivotal states in the SAHPC:

- Zimbabwe is directly impacted because the more variable flow of the Zambezi immediately translates into reduced hydropower capacity at Kariba. More importantly, the viability of Batoka Gorge (Knight Piésold & Lahmeyer International 1993) is likely to be in jeopardy (Fig. 18). The thermal energy based on the Hwange coalfields is also impacted, plunging Zimbabwe into a new energy crisis. Zimbabwe starts to look east to Mozambique for gas; and north to the DRC for both geothermal energy along the Great Rift Valley and hydropower from Grand Inga in the west of the country (see Fig. 19a, b).
- South Africa is hard hit as a result of the reduced yield from the Incomati and Maputo basins, with an immediate negative impact on the Mpumalanga-based Eskom operations. In similar vein the newly developed Eskom thermal energy plant in the Limpopo basin is placed at risk because of deteriorating water quality arising from return flows out of Gauteng, and Acid Mine Drainage from coal mining operations. South Africa is forced to revaluate the nuclear option, but growing public anger causes the government to look northwards towards the gas fields of Mozambique, and to the periodic surplus power available when the Zambezi is flowing strongly. Investment into North-South transmission infrastructure therefore grows, with a new corridor opening up for road, rail and energy pipelines direct from northern Mozambique into South Africa. The status of Mozambique is elevated as a regional power so economic growth in that country starts to hit double digit rates. At the same time South Africa increasingly looks towards the DRC for, hydropower from Inga in the west (see

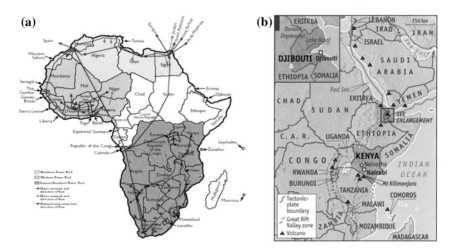

Fig. 19 a The Pan African energy grid showing major corridors that could become drivers of infrastructure investment. **b** Potential geothermal energy sources along the Great Rift Valley

Fig. 19). Given the sheer scale of this development, it lags behind the Mozambique transaction, but is still seriously considered.

- Botswana has invested into thermal energy based in the east of the country, but this is placed at risk because of water insecurity. The government therefore looks north towards the DRC, and the Zambezi and invests into the infrastructure corridor across the Zambezi River as Kasane. Road, rail and energy infrastructure upgrades in Zambia are routed to Botswana. Given the known stance of the Government of Botswana, they leverage this advantage with South Africa in order to form a joint venture between the two countries for the development of this infrastructure corridor (see Fig. 19a). Significant development thus takes place in Botswana, between Botswana and South Africa, and via Zambia into the DRC. Botswana invests wisely and finds water, energy and food security at regional level along this major infrastructure corridor.
- Namibia is the first to be plunged into crisis as the north of the country desiccates, resulting in a humanitarian disaster in the Cuvelai basin. The planned energy projects along the Kunene downstream of Calueque are all placed in jeopardy as major financial institutions call for a revaluation of the viability of those projects. Attention shifts to the offshore gas fields, but also towards the north where the logical first stop is a hydropower joint venture with Angola in some of the westerly flowing rivers of that country. The Namibian government is also aware of the Grand Inga scheme, so they apply their minds to the strategic implications of opening an infrastructure corridor to South Africa via Angola from Inga (see Fig. 19a). They leverage advantage from this position to offset the crisis unfolding in the north of the country.

9 Implications for Infrastructure Development

There is a general low level of infrastructural development in the SADC region, specifically with respect to water, transport and energy. More significantly, there is a coincidence between a growing need for water by the four pivotal states of the SAHPC, and the growing need for infrastructural development within the region that various donors can contribute to. A unique set of opportunities arise in the following:

- Aquifer storage and recovery (ASR), also known as managed aquifer recharge is an emerging technology that is increasingly being mainstreamed in water-constrained parts of the world (Moore et al. 2011). This type of project is ideal for a targeted, small scale infrastructure development project, specifically where the technology is being piloted as a demonstration for sceptics (Tredoux et al. 2002). Some of these are transboundary systems which have a different set of opportunities and challenges (Davies et al. 2012; Scheumann and Herrfahrdt-Pähle 2008). It is even possible to engineer such systems, specifically if mine rehabilitation involving an open pit is involved (Turton and Botha

2013). The significance of such systems is that they are very efficient because they reduce the evaporation losses associated with open dams in areas of high evaporative demand (Hut et al. 2008).

- It is clear that the region is facing a major challenge in the energy field, to the extent that the quest for viable solutions could potentially be a game changer for those directly involved. There is an opportunity to assist with the mapping of new energy sources, specifically those involving low grade heat sources such as those found along the southern-most extremity of the Great Rift Valley as it intersects the Caprivi Strip, and trading in the difference between average and firm energy to the south. These smaller resources are capable of being developed for localized small scale energy conversion, using technology developed in Europe and piloted on one of the small Greek islands. The larger resources are found further north and are already being exploited (Economist 2008; Mbuthi and Yuko 2005).
- Water infrastructure, specifically suited to small-scale agriculture focussed on poverty eradication and climate change resilience-building is a potentially low hanging fruit, but would require significant replication to make any substantial impact on regional poverty. Of particular interest are low maintenance concepts such as rainwater harvesting, specifically where this might be linked to the production of food or the recharging of localized aquifers. Water harvesting and reuse is also likely to be of increased importance, specifically in the rural areas of the pivotal states as water constraints inhibit substantial development in the future. More importantly, corridors that are suitable for multiple forms of infrastructure, such as those used in Botswana, where electricity, roads, railway lines and pipelines all run parallel in a well-designed system that is easy to service and maintain.
- The water stewardship roles of the private sector, expanding Corporate Social Responsibility roles, and an increase in the numbers of outgrower schemes could serve as a key driver of rural development. This may help push agricultural production northwards making better use of the regions green water resources—particularly in the face of poor runoff coefficients.
- Transport infrastructure is going to be a key element in regional integration, specifically under conditions identified in the scenario presented in this paper. There are likely to be three major North-South transport corridors, each capable of multiple use for electricity and potentially even water or gas pipelines. The Eastern Corridor is likely to go from South Africa up to the northern portion of Mozambique. The Central Corridor is likely to go from Botswana to the DRC via the bridge over the Zambezi located at Kasane. The Western Corridor is likely to go from South Africa, via Namibia and Angola to the DRC, driven mostly by the need to service the electricity grid should Grand Inga come online. This transport infrastructure will need to be serviced by rapid border crossing facilities, specifically if fresh food is to make it to the markets in the south in an unspoilt manner.
- Climate change is likely to create a new set of health-related opportunities and threats (Bradshaw et al. 2003). Opportunities will present as increased

penetration into the rural areas arising from infrastructure development (energy, transport, pipelines). Threats are likely to arise from a change in disease vectors such as mosquitoes, along with the types of problem associated with DDT use (Aneck-Hahn et al. 2007; Bornman et al. 2005, 2009). Another specific category of threat will be associated with the exposure of persons with compromised immune systems to microcystin toxins in hypertrophic waters (Codd et al. 1999; Doyle 1991; Falconer 1998, 2005; Humpage et al. 2000; Oberholster et al. 2004, 2005).

- Food production is likely to undergo a major change in the SADC region in the next few decades. One manifestation will probably be the relocation of major production centres to the wetter north, more specifically in the DRC, Angola, Zambia and northern Mozambique. While this is likely to favour large-scale commercial farming, there is a strong need to foster a parallel system centred on small-scale farmers, possibly organized around cooperative structures capable of servicing the market needs in the south where better prices are likely to be found. This represents a specific targeted opportunity for early assessment and intervention. Another area is that associated with the monitoring of water quality for crop production, specifically from hypertrophic water where microcystin might be a limiting factor (Abe et al. 1996).

10 King's Demographic Trap as a Concept

While King's Demographic Trap is contested by some (Sen 1989, 2000), it remains a useful tool in understanding how water might become both a national and regional risk. In the professional opinion of the author, this should not be dismissed without a thorough interrogation in a SADC context. It must be noted that the pathway to King's Demographic Trap is not a linear one as shown in Fig. 20, being driven by a combination of growing population pressure, increased generation of waste and manifesting as accelerated water pollution—the exact type that is becoming evident in South Africa as eutrophic ecosystems.

Text Box 1 King's Demographic Trap
The survival of a population depends ultimately on a sustainable supply of essential resources, particularly fresh water and food. If these are not available in sufficient quantities to sustain the people living in a nation or region, the population has exceeded the carrying capacity of that nation or region. Both populations and supplies of fresh water and food are dynamic, not static. Usually, in most nations, there is a positive balance—the nation or region either has, or can afford to import, a sufficient supply of fresh water and food to enable all currently living to survive, with enough left over to allow for natural population increase. However, sometimes the rate of increase of a

nation's or region's population is greater than the capacity of the local or regional ecosystems to produce the food that is necessary for all to survive, and there are no financial resources to import these necessities for survival. Moreover, natural or manmade disasters can tip the balance by disrupting food supplies. A population that has exceeded the national or regional carrying capacity is said to be caught in a demographic trap (King and Elliot 1993). Such a population must migrate out of the region, or it will starve unless it receives food aid. Another possible consequence may be violent armed conflict if the demographically trapped population encroaches on the territory of neighbouring nations who regard them as unwelcome intruders. The concept of the demographic trap first appeared in the annual report of the Worldwatch Institute in 1987. It was discussed at a major World Health Organization (WHO) conference in 1988, and has been much discussed since then.

Source: Encyclopaedia of Public Health (http://www.answers.com/topic/demographic-trap).

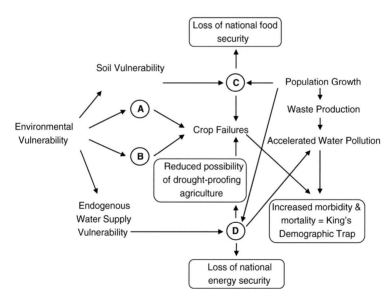

Fig. 20 Schematic representation of the main sovereign risks along with different pathways to environmental vulnerability (adapted from Falkenmark 1994: 20)

In order to understand the different pathways that sovereign risks might arise from water, Fig. 20 makes use of four distinct modes of water scarcity as defined by Falkenmark (1993, 1994). The ramifications of these are as follows:

- **Mode "A"**—a lack of "green water" (water used in the production of biomass arising from natural precipitation). This can be thought of in its most simple form as a short growing season arising from rainfall that is late or erratic. Technically this is what happens where localized rain falls but a major drought prevails in the broader region. (sometimes referred to as meteorological droughts).
- **Mode "B"**—intermittent drought such as that which occurs over much of the SADC region. This is the type of drought that would arise from an *el Nino* event as shown in Fig. 3.
- **Mode "C"**—manmade desiccation of the landscape arising from poor agricultural practices. Typically this occurs in areas with a high water crowding index (WCI) value.
- **Mode "D"**—a lack of "blue water" (water that has been trapped in dams and engineered systems) both endogenous (occurring inside the country) and exogenous (arising outside the country but flowing into it via aquifers or rivers). This is where eutrophication would feature, because it impacts water in dams, rendering it unfit for purpose even if it is available in a volumetric sense of that word. (sometimes referred to as hydrological droughts).

From the conceptual model shown in Fig. 20, four significant and distinct risks arise from water. The ramifications of these is shown in a rounded box and can be described as follows:

- Reduced possibility of drought-proofing agriculture. This is driven mostly by **Mode "D"** and is most likely to manifest in areas with a high WCI. A sub-set of this is human health impacts, such as those now manifesting as intersexed babies in areas with a high WCI superimposed onto areas with endemic malaria (typically Limpopo).[3] Loss of national energy security. This is also associated with **Mode "D"** risk and is driven by vulnerability to national water resources as a result of the overall demand for water (most notably for food production) exceeding sustainable supply and thus becoming a constraint to hydrothermal energy production.

---

[3]A study by Bornman et al. (2009) has shown that in a test sample of 3310 new-born male babies in the Limpopo area, 357 (10.8 % of the total sample) had various forms of urogenital birth defects. A statistical analysis of this sample revealed that a mother living in an area that was sprayed with DDT from 1995 to 2003 had a 33 % increased chance of giving birth to a male child with a urogenital birth defect. By being a homemaker and thus unemployed further increased the risks by 41 %. There are other examples such as impaired semen quality (Aneck-Hahn et al. 2007) and urogenital defects in males (Bornman et al. 2005). This has significant implications if these findings are accurate, because they become yet another driver of what is already manifesting as increased vulnerability of subsistence agriculture to **Mode "D"** risk.

- <u>Loss of national food security</u>. This is driven by **Mode "C"** risk, arising from a high WCI in combination with increased soil vulnerability.
- <u>Increased morbidity and mortality arising from King's Demographic Trap</u>. This is driven by population growth in water-stressed areas (manifesting first as a high WCI) combining later with **Mode "D"** risk as systems start to exceed design parameters and fail.

11 Conclusion

This chapter has attempted to generate a set of scenarios, based on three core assumptions: that climate change is likely to make the SADC region both hotter and drier; that a hydropolitical complex exists in the region in which the tendency for inter-state behaviour is to converge around cooperation rather than competition; and that sovereignty can be managed in a way that need not necessarily mean a direct challenge to the supremacy of the state. If correct, then the countries located in the more economically diverse south are likely to be increasingly unable to create national self-sufficiency in water, energy and food. This will drive a hard set of decisions that will have to be made. Inevitably however, the solutions to these three vexing problems are likely to be found at regional rather than national level. The better-watered North will increasingly become the regional breadbasket, with the DRC, Angola, Zambia and northern Mozambique emerging as major players. If one overlays the energy crisis onto the food challenge, then we start to see the need for three major infrastructure corridors—the eastern, central and western—capable of servicing the water, energy and food needs of the south from the north. Intra-regional trade is likely to change in a significant manner, driving the need for liberalization and faster border crossings at selected choke points. Policy integration will therefore be necessary and can be considered soft infrastructure.

References

Abe T, Lawson T, Weyers JDB, Codd GA (1996) Microcystin-LR inhibits photosynthesis of *Phaseolus vulgaris* primary leaves: implications for current spray irrigation practice. New Phytol 133:651–658

ADB (2000) Atlas of Africa. Les Éditions du Jaguar, Paris. ISBN 2-86950-329-6

Africa (1984) SADCC moves to tackle drought. Africa, vol 153, pp 74–75

Allan JA (2011) Virtual water: tackling the threat to our planet's most precious resource. I.B. Tauris, London

Aneck-Hahn NH, Schulenburg GW, Bornman MS, Farias P, de Jager C (2007) Impaired semen quality associated with environmental DDT exposure in young men living in a malaria area in the Limpopo Province, South Africa. J Androl 28(3):423–434

Ashton PJ, Turton AR (2008) Water and security in sub-Saharan Africa: emerging concepts and their implications for effective water resource management in the southern African region. In: Brauch H-G, Grin J, Mesjasz C, Krummenacher H, Behera NC, Chourou B, Spring UO, Liotta PH, Kameri-Mbote P (eds) Facing global environmental change: environmental, human, energy, food, health and water security concepts—volume IV. Springer, Berlin, pp 665–678

Basson MS (1995) South African water transfer schemes and their impact on the Southern African region. In: Matiza T, Craft S, Dale P (eds) Water resource use in the Zambezi Basin. Proceedings of a workshop held in Kasane, Botswana. IUCN, Gland, Switzerland, 28 Apr–2 May 1993

Baynham S (1989) SADCC security issues. Afr Insight 19(2):88–95

Bornman MS, Delport R, Becker P, Risenga SS, de Jager CP (2005) Urogenital birth defects in neonates from a high-risk malaria area in Limpopo Province, South Africa. Epidemiology 5 (16):S126–S127

Bornman MS, de Jager CP, Worku Z, Farias P, Reif S (2009) DDT and urogenital malformations in newborn boys in a Malaria Area. Br J Urol (BJU) Int

Bradshaw D, Groenewald P, Laubscher R, Nannan N, Nojilana B, Norman B, Pieterse D, Schneider M (2003) Initial burden of disease estimates for South Africa, 2000. South African Medical Research Council, Cape Town

Cavé LC, Beekman HE, Weaver JMC (2003) Impact of climate change on groundwater recharge estimation. In: Xu Y, Beekman HE (eds) Groundwater recharge estimation in Southern Africa. UNESCO IHP series No. 64. UNESCO, Paris, pp 189–197. ISBN 92-9220-000-3

Cobbing JE, Hobbs PJ, Meyer R, Davies J (2008) A critical overview of transboundary aquifers shared by South Africa. Hydrogeol J 16(6):1207–1214

Codd GA, Metcalf JS, Beattie KA (1999) Retention of *Microcystis aeruginosa* and microcystin by salad lettuce after spray irrigation with water containing cyanobacteria. Toxicon 37:1181–1185

Davies J, Robins NS, Farr J, Sorenson J, Beetlestone P, Cobbing JE (2012) Identifying transboundary aquifers suitable for international resource management in the SADC region of Southern Africa. J Hydrogeol

Doyle P (1991) The impact of AIDS on the South African population. AIDS in South Africa: the demographics and economic implications. Centre for Health Policy, University of the Witwatersrand, Johannesburg, South Africa

Earle A, Turton AR (2003) The virtual water trade amongst countries of the SADC. In: Hoekstra A (ed) Virtual water trade: proceedings of the international experts meeting on virtual water trade. Delft, the Netherlands. Research Report Series No. 12. IHE, Delft, pp 183–200, 12–13 Dec 2002

Economist (2008) Geothermal power in Africa: continental Rift: a hot new proposal for ending electricity shortages. Available online at http://www.economist.com/node/12821590?story_id=12821590

Falconer IR (1998) Algal toxins and human health. In: Hrubec J (ed) Handbook of environmental chemistry, volume 5 (Part C). Springer, Berlin, pp 53–82

Falconer IR (2005) Cyanobacterial toxins of drinking water supplies: cylindrospermopsins and Microcystins. CRC Press, Florida, 279p

Falkenmark M (1993) Landscape as life support provider: water-related limitation, population summit of the World's Scientific Academies, New Delhi, October 1993. In: Graham-Smith F (ed) Population—the complex reality. Royal Society, London

Falkenmark M (1994) The dangerous spiral: near-future risks for water related eco-conflicts. In: ICRC. 1994. Water and war. Symposium on water in armed conflicts. Montreux. International Committee of the Red Cross (ICRC), Geneva, pp 10–28, 21–23 November 1994

Heyns P (2002) Interbasin transfer of water between SADC countries: a development challenge for the future. In: Turton AR, Henwood R (eds) Hydropolitics in the developing world: a southern African perspective. African Water Issues Research Unit (AWIRU), Pretoria, pp 157–176

Hoekstra AY, Hung PQ (2002) Virtual water trade: a quantification of virtual water flows between nations in relation to international crop trade. Research Report No. 11. UNESCO IHE

Homer-Dixon TF (1994) Environmental scarcities and violent conflict: evidence from cases. Int Secur 19(1):5–40

Homer-Dixon TF (1996) Environmental scarcity, mass violence and the limits to ingenuity. Curr Hist 95(1):359–365

Humpage AR, Hardy SJ, Moore EJ, Froscio SM, Falconer IR (2000) Microcystins (cyanobacterial toxins) in drinking water enhance the growth of aberrant crypt foci in the colon. J Toxicol Environ Health 61:101–111

Hut R, Ertsen M, Joeman N, Vergeer N, Winsemius H, Giesen N (2008) Effects of sand storage dams on groundwater levels with examples from Kenya. Phys Chem Earth 33(1–2):56–66

IGRAC (2009) Transboundary aquifers of the World. Available online at http://www.un-igrac.org/dynamics/modules/SFIL0100/view.php?fil_Id=121

King M, Elliot C (1993) Legitimate double think. Lancet 341:669–672

Knight Piésold & Lahmeyer International (1993) Batoka Gorge hydroelectric scheme feasibility report. Batoka Joint Venture Consultants. Zambezi River Authority

MacDonald M, Partners (1990) Sub Saharan Africa hydrological assessment: SADCC countries: regional report

Maupin A (2010) L'espace Hydropolitique de l'Afrique Australe. Le Risque Hydropolitique dans les Politiques de Gestion de l'Eau des Bassin Transfrontaliers. PhD Thesis, University Michel de Montaigne, Bordeaux 3, France

Mbuthi P, Yuko D (2005) Potential renewable energy technologies in Kenya's electricity supply: a review of geothermal and cogeneration technologies. Occasional Paper No. 26. African Energy Research Policy Network (AFREPREN/FWD), Nairobi

Mekonnen MM, Hoekstra AY (2011) The green, blue and grey water footprint of crops and derived crop products. Hydrol Earth Syst Sci 15(5):1577–1600

Moore JM, Marley R, Schuster S (2011) Managed aquifer recharge. Daniel B. Stephens and Associates (Inc), USA

Nielsson G (1990) The parallel national action process. In: Groom AJR, Taylor P (eds) Frameworks for international cooperation. Pinter Publishers, London, pp 78–108

Nye J (1971) Peace in parts: integration and conflict in regional organisation. Little, Brown & Co., Boston

O'Keeffe J, Uys M, Bruton MN (1992) Freshwater systems. In: Fuggle RF, Rabie MA (eds) Environmental management in South Africa. Juta & Co., Johannesburg, pp 277–315

Oberholster PJ, Ashton PJ (2008) State of the nation report: an overview of the current status of water quality and eutrophication in South African rivers and reservoirs. Parliamentary grant deliverable. Council for Scientific and Industrial Research (CSIR), Pretoria

Oberholster PJ, Botha A-M, Grobbelaar JU (2004) Microcystis aeruginosa: source of toxic microcystins in drinking water. Afr J Biotechnol 3:159–168

Oberholster PJ, Botha AM, Cloete TE (2005) An overview of toxic freshwater cyanobacteria in South Africa with special reference to rosk, impact and detection by molecular marker tools. Biochem 17:57–71

Oberholster PJ, Cloete TE, van Ginkel C, Botha A-M, Ashton PJ (2008) The use of remote sensing and molecular markers as early warning indicators of the development of cyanobacterial hyperscum crust and microcystin-producing genotypes in the hypertrophic Lake Hartebeespoort. Council for Scientific and Industrial Research (CSIR), South Africa. Pretoria

Pallett J, Heyns P, Falkenmark M, Lundqvist J, Seeley M, Hydén L, Bethune S, Drangert J, Kemper K (1997) Sharing water in Southern Africa. Desert Research Foundation of Namibia (DRFN), Windhoek

Percival V, Homer-Dixon T (2001) The case of South Africa. In: Diehl PF, Gleditsch NP (eds) Environmental conflict. Westview Press, Boulder, pp 13–35

Ramoeli P (2002) SADC protocol on shared watercourses: its history and current status. In: Turton AR, Henwood R (eds) Hydropolitics in the developing world: a southern African perspective. African Water Issues Research Unit (AWIRU), Pretoria, pp 105–111

Scheumann W, Herrfahrdt-Pähle E (eds) (2008) Conceptualizing cooperation for Africa's transboundary aquifer systems. German Federal Ministry for Economic Cooperation and Development (BMZ), Bonn

Scholes RJ (1998) The impact of global change on South Africa: facts and fallacies. Tydskrif vir Skoonlug 10(1):23–24

Scholes RJ, Biggs R (2004) Ecosystem services in Southern Africa: a regional assessment. CSIR, Pretoria

Sen A (1989) On ethics and economics. Blackwell, Oxford

Sen A (2000) Development as freedom. Doubleday, New York

Snaddon CD, Davies BR, Wishart MJ (1998) A Global Overview of Inter-Basin Water Transfer Schemes, with an Appriasal of their Ecological, Socio-Economic and Socio-Political Implications, and Recommendations for their Management. Water Research Commission Report No. TT 120/00. Pretoria: Water Research Commission

TPTC (2008) Groundwater development options and conjunctive use. Consultancy for the integrated comprehensive study of the water resources of the Maputo River Basin— Mozambique, Swaziland and South Africa. Maputo: Tripartite Permanent Technical Committee (TPTC) Republic of Mozambique, Republic of South Africa and Kingdom of Swaziland

Tredoux G, Murray EC, Cavé LC (2002) Infiltration basins and other recharge systems in Southern Africa. In: Tuinhof A, Heederik JP (eds) Management of aquifer recharge and subsurface storage: making better use of our largest reservoir. NNC-IAH Publication No. 4. Netherlands National Committee of the International Association of Hydrogeologists, Wageningen

Turton AR (2001a) A hydropolitical security complex and its relevance to SADC. Conflict Trends Issue 1/2001:21–23. ISSN 1561-9818

Turton AR (2001b) Hydropolitics and security complex theory: an African perspective. Paper presented at the 4th Pan-European international relations conference, University of Kent, Canterbury (UK), 8–10 Sept 2001

Turton AR (2002) Water and State Sovereignty: the hydropolitical challenge for states in Arid regions. In: Wolf A (ed) Conflict prevention and resolution in water systems. Edward Elgar, Cheltenham, pp 516–533

Turton AR (2003) Transboundary rivers: a strategic issue in SADC. SADC Barometer 3:18–19

Turton AR (2004) The evolution of water management institutions in select Southern African international river basins. In: Biswas AK, Unver O, Tortajada C (eds) Water as a focus for regional development. Oxford University Press, London, pp 251–289

Turton AR (2007) The hydropolitics of cooperation: South Africa during the cold war. In: Grover VE (ed) Water: a source of conflict or cooperation? Science Publishers, Enfield, NH, pp 125–143. ISBN 978-1-57808-511-8. Formerly CSIR Report No: ENV-P-R 2005-008

Turton AR (2008) Discussion paper on parallel national action (PNA) as a potential model for policy harmonization in the SADC region. GTZ Contract No. 026/08. CSIR Report No. CSIR/NRE/WR/ER/2008/0108/C. SADC Secretariat, Gaborone

Turton AR, Ashton PJ (2008) Basin Closure and issues of scale: the Southern African hydropolitical complex. Int J Water Resour Dev 24(2):305–318

Turton AR, Botha FS (2013) Anthropocenic aquifer: new thinking. In: Eslamien S (ed) Handbook for engineering hydrology (volume 3): environmental hydrology and water management. Francis & Taylor, London (Chapter 59)

Turton AR, Warner J (2002) Exploring the population/water resources nexus in the developing world. In: Dabelko GD (ed) Finding the source: the linkage between population and water. Environmental change and security project (ECSP). Woodrow Wilson Centre, Washington, DC, pp 52–81

Turton AR, Moodley S, Goldblatt M, Meissner R (2000) An analysis of the role of virtual water in Southern Africa in meeting water scarcity: an applied research and capacity building project. Group for Environmental Monitoring (GEM), Johannesburg

Turton AR, Schultz C, Buckle H, Kgomongoe M, Malungani T, Drackner M (2006a) Gold, scorched earth and water: the hydropolitics of Johannesburg. Water Resour Dev 22(2):313–335

Turton AR, Patrick MJ, Cobbing J, Julien F (2006b) The challenges of groundwater in Southern Africa. Environmental change and security program navigating peace, vol 2. Woodrow Wilson International Centre for Scholars, Washington

Turton AR, Ashton PJ, Jacobs I (2008) The management of shared water resources in Southern Africa. CSIR Report No. CSIR/NRE/WR/ER/2008/0400/C. IMIS Contract No. 2009UNA073263853111. United Nations Economic Commission for Africa-Southern Africa (UNECA-SA), Lusaka

Ueno Y, Nagata S, Tsutsumi T, Hasegawa A, Watanabe MF, Park HD, Chen GC, Yu S (1996) Detection of microcystins in blue-green algae hepatotoxin in drinking water sampled in Haimen and Fusui, endemic areas of primary liver cancer in China, by highly sensitive immunoassay. Carcinogenesis 17:1317–1321

UNEP (2002a) Vital water graphics: an overview of the state of the world's fresh and marine waters. United Nations Environment Program (UNEP): Nairobi

UNEP (2002b) Atlas of international freshwater agreements. United Nations Environment Program; and Corvallis: Oregon State University, Nairobi

Wolf AT (2006) Hydropolitical vulnerability and resilience: series introduction. In: Wolf AT (ed) Hydropolitical vulnerability and resilience along international waters: Africa. United Nations Environment Program (UNEP), Nairobi, pp 3–17

World Bank (1992) SADCC energy project AAA 3.8. Regional generation of transmission capacities including interregional pricing policies. Phase II. Hydrology Report

Virtual Water and the Nexus in National Development Planning

Mike Muller

Abstract Because of their great variety of uses and impacts, the development and management of water resources has to be coordinated with the needs of users. Hydro-centric approaches such as 'Dublin' Integrated Water Resource Management (IWRM) convene stakeholders to water-focused processes on a river basin scale and emphasise environmental conservation rather than resource development. Hydro-supported processes work at the scale of political units and focus on 'problem-sheds', demand centres and supply systems, rather than river basins and develop multi-purpose rather than single purpose responses. As mandated at the UN's Mar del Plata water conference, they seek integration with national development strategies. The evidence suggests that hydro-supportive processes are more effective in coordinating water management with other sectors because they operate at common political and administrative scales. Concepts such as "Virtual Water" and the "water-food-energy nexus" may usefully inform national and regional development planning by helping to identify inter-sectoral trade-offs and synergies. But they are unlikely to provide the basis for national policies on which regional cooperation and action depend, given the many other factors that have to be considered.

1 Introduction and Background

Text Box 1 Complementary Endowments Offer Opportunities

Complementary endowments offer opportunities

Minister Trevor Manuel, chairman of South Africa's National Planning Commission and champion of the SADC/Nepad North-South Corridor project, has highlighted the opportunities offered by greater regional cooperation:

M. Muller (✉)
Wits University's School of Governance, Johannesburg, South Africa
e-mail: mikemuller1949@gmail.com

© Springer International Publishing Switzerland 2016   87
A. Entholzner and C. Reeve (eds.), *Building Climate Resilience through Virtual Water and Nexus Thinking in the Southern African Development Community*, Springer Water, DOI 10.1007/978-3-319-28464-4_4

As we imagine different futures for our different countries, we should also have the courage to imagine ourselves working together as a single region. If we do that, we find that the balance of our endowments looks a little different. If we combine our access to capital as a region, with the diversity of human resources that we have, the independence dividend that is now maturing in the region, with our extensive natural resources [...] a completely different set of opportunities would arise. And while we would still have large numbers of relatively unskilled people, they would have far wider opportunities than if we simply worked as individual countries. (Manuel 2011).

Water is a factor of greater or lesser importance in many economic and social sectors and its management (as a resource) and provision (as a service) are often considered to be economic and social "sectors" in their own right. To the extent that there is a generic goal for water resource management, it is to achieve water security for society, defined as *"the availability of an acceptable quantity and quality of water for health, livelihoods, ecosystems, and production, coupled with an acceptable level of water-related risks to people, environments, and economies."* (Grey and Sadoff 2007: 546).

The management of water as a renewable, "common-pool" natural resource whose presence is both variable and unpredictable, poses many challenges. Although freshwater is an important factor of production in many sectors, it does not need to be produced; the resource must rather be developed and managed. The immediate concern of "user" sectors is usually the quantity of water available to them. However, the need to maintain water resource quality both to sustain desired environmental conditions as well as to avoid prejudice to other users becomes increasingly important as levels of use increase. In many countries, management of flood impacts is also an essential function.

From a development policy and strategy perspective, water is a contextual resource endowment rather than a driving force. While, historically, early agricultural civilisations may have developed into "hydraulic societies", the linkages between water and societal economic and social development have weakened as our ability to manage water to meet development needs has increased. Outside of agriculture and hydropower, water availability is seldom a dominant determinant of the location of economic activity and water resource development and management is guided by demand rather than used to catalyse activity through supply. The dominant approach to water management has been to get the water to where it is needed, rather than to develop where the water is available—particularly in southern Africa.

The nature of management activities is often complex since it has to deal with extreme variability and uncertainty as well as the geographic location of the resource which is often not available in adequate quantities where it is needed without infrastructure investments. As use intensifies, there is often competition between users for access to limited supplies and a system has to be established that guides the allocation of what is usually considered to be a public resource. This process has to take account of changing social and economic priorities and preferences. A further challenge is to take account of the need to sustain the resource and its underlying biodiversity and to reflect the environmental preferences and priorities of society.

While there are always likely to be infrastructure solutions to water availability and variability, these may become increasingly economically and environmentally expensive. Regional differences in water endowments may similarly require the transfer of water over longer distances between basins and nations or demand other responses.

A final contextual issue is the challenge of climate variability and change. There has been extensive discussion about the potential impacts of climate change on water resources, with warnings that it may amplify the destructive impacts of both flooding and droughts. There is however a widely held view amongst practitioners that current climate variability already requires a structured management response, which many communities and countries are still not able to provide. The preferred strategy for climate change is thus to build community, country and regional resilience by building the capacity to address current climate variability (Sadoff and Muller 2008).

There has been extensive discussion about the potential impacts of climate change on water resources, with warnings that it may amplify the destructive impacts of both flooding and droughts. There is however a widely held view amongst practitioners that current climate variability already requires a structured management response, which many communities and countries are stil not able to provide. The preferred strategy for climate change is thus to build community, country and regional resilience by building the capacity to address current climate variability (Sadoff and Muller 2008).

2 Water Resource Planning Is Contested Terrain

If water-related development decisions are to be influenced, it is necessary to understand the associated decision-making processes about water resource development, management and use and how diverse water-using sectors are engaged in these. It is also important to recognise that this is a contested terrain; a full description of the recent evolution of different approaches is beyond the scope of this paper (see Muller 2015 for more detail).

Because of its multi-sectoral use and impacts, water resource development and management has to be closely coordinated with the needs of users and water resource management institutions should be able to inform user sectors of the opportunities and constraints that water may pose for their activities. Two broad macro-approaches to introducing water issues into national policy can be distinguished, hydro-centric and hydro-supported.

A variety of "hydro-centric" planning processes have been promoted, often by environmental conservation interest groups. These are characterised by an attempt to "put water at the centre of development", to make the physical boundaries of river basins the primary scale at which water is planned (e.g. European Water Framework Directive). In particular, they seek to resolve development trade-offs between different sectors in forums established by water sector institutions. An earlier generation of hydro-centric processes (the USA's TVA scheme is the flag-ship for these) sought to stimulate development through investments in water infrastructure. These had mixed results and, currently, the emphasis of hydro-centric results is to protect rather than to develop the resource.

Hydro-centric approaches are particularly difficult to apply in transboundary river systems since they require water resources that are shared between nations to be managed jointly by an over-arching river basin organisation. It is difficult for such institutions to negotiate trade-offs between riparian states where the benefits accrue to one state and the costs are incurred in another. 'Benefit sharing', while often touted as the principle that should govern transboundary management is hampered by the complexities of agreeing a reasonable and equitable share of those benefits, particularly where these accrue to and from different sectors of economies and societies which do not have adequate voice in the management process.

Hydro-supported processes are those in which the development and management of the resource is guided by agencies which are part of a wider family of political and administrative institutions. These are driven primarily by user requirements and such user-led approaches are typical of most rapidly developing countries. The most obvious user-requirement in these cases is for adequate quantity and reliability of water supplies. Regulation of resource quality impacts is more difficult. While individual users can be required by water managers to treat waste discharges to certain standards to protect other users, the management of diffuse impacts must involve other sectors. So "diffuse" pollution caused by agricultural practices needs to be regulated in cooperation with the relevant agricultural authorities through formal governmental coordination processes.

Similarly, at regional level, the implications of differences between national water endowments will have to be addressed as part of overall economic management. So the viability of the large and costly intra-regional water transfers mooted by some authors as a solution to SADC wide variability in water availability will be informed by the economic perspectives of the user sectors rather than by water managers.

At a global policy level, the water-sector has, in recent decades (1992–present), been encouraged to follow what are effectively hydro-centric processes in which the conservation and even "preservation" of the resource is prioritised (IWRM, river basin planning) but these processes have had relatively limited impacts and out-comes. An earlier (1930–1990) set of hydro-centric approaches focused on the promotion of large water resource infrastructure programmes intended to catalyse economic and social development. Some of these are considered to have been successful (TVA, 3 Gorges) while others have had more mixed results (Kariba, which has not seen significant irrigation development) and some are widely regarded as failures (Mekong, where the instability after the Vietnam war paralysed, until recently, the planned infrastructure developments). While large hydro-centric resource programmes have often captured the imagination of both water sector managers and politicians, it is suggested that, in terms of economic and social impact, it is the hydro-supported processes that have had the greatest impact although, because this is "indirect", it is less visible. If water-related development decisions are to be influenced today, it is useful to understand and track the recent evolution of these approaches.

Because of the contribution that water resources and their management make to so many different areas of human social and economic activity, it has long been suggested that water resource development and management should be addressed as part of overall national development strategy and planning. This was explicitly stated in the 1977 UN Conference on Water at Mar del Plata which sought to identify and recommend the actions needed for the "accelerated development and orderly administration of water resources". Its Action Plan placed considerable focus on the need for a more coherent approach, emphasising the need for a

> .... shift from single-purpose to multipurpose water resources development as the degree of development of water resources and water use in river basins increases, with a view, *inter alia*, to optimizing the investments for planned water-use schemes. In particular, the construction of new works should be preceded by a detailed study of the agricultural, industrial, municipal and hydropower needs of the area concerned. [...] This analysis would take into account the economic and social evolution of the basin and be as comprehensive as possible; it would include such elements as time horizon and territorial extent, and take into account interactions between the national economy and regional development, and linkages between different decision-making levels. (UN 1977: para 41).

To achieve this, it was recommended that the management of water resources should be effectively integrated and explicitly proposed that this should be through the mechanism of national development planning:

> Each country should formulate and keep under review a general statement of policy in relation to the use, management and conservation of water, as a framework for planning and implementing specific programmes and measures for efficient operation of schemes. National development plans and policies should specify the main objectives of water-use policy, which should in turn be translated into guidelines and strategies, subdivided, as far as possible, into programmes for the integrated management of the resource. (UN 1977: para 43).

This theme was taken up again 15 years later at the UN Summit on Sustainable Development in Rio de Janeiro. The Action Plan prepared there, Agenda 21, states that:

> The holistic management of freshwater as a finite and vulnerable resource, and the integration of sectoral water plans and programmes within the framework of national economic and social policy, are of paramount importance for action in the 1990s and beyond. (Chap. 18)

However, divides emerged between the developed countries that wanted to emphasise environmental sustainability and developing countries that sought greater emphasis on their economic and social development.

This is illustrated by the way in which the currently dominant hydro-centric approach was outlined in the final statement of a preparatory meeting held in Dublin before the Rio Conference. It focuses exclusively on basin level planning (its only mention of national development plans is in relation to training needs).

> The most appropriate geographical entity for the planning and management of water resources is the river basin, including surface and groundwater. (Dublin 1992).

Water sector planning processes were seen as essential to the resolution of water conflicts. The Dublin statement also explicitly gave priority to environmental objectives:

> Integrated management of river basins provides the opportunity to safeguard aquatic ecosystems, and make their benefits available to society on a sustainable basis.

Many of the key proposals made in Dublin were rejected by the Rio Conference. Aside from its emphasis on economic instruments over social objectives, recommendations from Dublin that were not taken up in Agenda 21 included: that river basins should be the unit of decision making; that stakeholders should participate fully in decisions; that future international meetings on water should be convened as multi-stakeholder fora in which governments would have the same role as business and NGOs. Nevertheless, the so-called "Dublin Principles" were widely adopted, particularly by donor countries in relation to their aid recipients.

One outcome of the Dublin Principles focus on environment, river basins and stakeholder participation was the convening of a World Commission on Dams. The Commission was dominated by anti-dam NGOs and its recommendations for reviews of alternatives to dam development and full prior consent by affected parties before development were widely regarded as unworkable. The result was that its report, in the words of one long-time observer of the water sector put it:

> … contributed to a concerted action by the developing countries which were forced to unite by the biased report which otherwise may not have happened. With a combined voice, they could tell developed countries who had already constructed most of their large dams, that infrastructure construction is important for their socio-economic development and that they need such structures to produce food, generate energy employment and income, provide basic services and improve the overall quality of life of their citizens (Biswas 2012).

One outcome was however that donor countries and agencies became very reluctant to finance large water infrastructure and, although this position has moderated somewhat, the negative attitudes are still in place as demonstrated by the fact that large hydropower dams are still not eligible for Clean Development Mechanism financing.

The approach inherent in the Dublin Principles was also reflected in the European Union's Water Framework Directive which was approved in 2000. This again focused on the environmental integrity of river basins, with basins as a unit of planning and full stakeholder participation. As described by the European Commission, the environmental requirements appear particularly onerous:

> …. ecological protection should apply to all waters: the central requirement of the Treaty is that the environment be protected to a high level in its entirety. […] the controls are specified as allowing only a slight departure from the biological community which would be expected in conditions of minimal anthropogenic impact. (EC WFD introductory note)

But European politicians refused to endorse proposals for river basin organisations to take responsibility for transboundary rivers—the requirement for "coordination" allowed most to carry on with business as usual although with additional reporting requirements. There were, nonetheless, requirements for aligning monitoring and reporting systems, to ensure 'good' status was not reported as 'fair' just over the border. Aspirations to re-establish natural conditions were considerably diluted and sufficient loopholes were left to give national governments extensive discretion—the Netherlands simply declared the majority of its watercourses to be artificial (Heavily Modified Water Bodies), which only need to achieve good chemical status. The requirement for stakeholder participation is also being questioned; some governments find that they can only comply by paying participants to attend meetings.

After 1992, two institutions (the Global Water Partnership (GWP) and World Water Council), which were established outside the UN system to give effect to the Dublin Principles (rather than Rio's Agenda 21), focused on this approach. The GWP and the Scandinavian governments that backed it took the lead in promoting Integrated Water Resource Management (IWRM) plans and elaborating how they should be produced. Although characterised as integrated approaches, they were conceived as water sector led initiatives.

> The promotion of catchment and river basin management is an acknowledgement that these are logical planning units for IWRM from a natural system perspective. Catchment and basin level management is not only important as a means of integrating land use and water issues, but is also critical in managing the relationships between quantity and quality and between upstream and downstream water interests (GWP 2000).

The consequence of this hydro-centric approach was to concentrate on water sector based instruments rather than effective coordination with broader social and economic development—and political—processes.

> .... in many cases stakeholders represent conflicting interests and their objectives concerning water resources management may substantially differ. To deal with such situations the IWRM should develop operational tools for conflict management and resolution as well as for the evaluation of trade-offs between different objectives, plans and actions.

In 2002, at the Johannesburg World Summit on Sustainable Development, after strong lobbying by European delegations, it was agreed that all countries should prepare IWRM plans by 2005. This marked a turning point since it subsequently became clear that the nature and purpose of these plans was unclear and was based on a poor understanding of how water resource related matters were managed in practice. While a number of developing countries were funded to prepare such plans, they have had little impact, not least because the guidelines for their preparation focused on institutions and management instruments and almost completely ignored the infrastructure needed in most countries to enable such institutions and instruments to operate.

In response to the overwhelming emphasis on process and institutions and the underwhelming practical outcomes it became clear that a different focus was required. One response, emerging from the World Bank, was to focus on the achievement of the practical goal of water security, "the availability of an acceptable quantity and quality of water for health, livelihoods, ecosystems and production, coupled with an acceptable level of water-related risks to people, environments and economies". The overarching strategy to achieve this was to invest in the institutions, information and infrastructure needed to achieve the goal.

At the same time, the World Bank sought to re-engage in infrastructure for water resource management.

A related but more specific response subsequently emerged from the business community, which recognised the need for practical outcomes to address the growing economic and social challenges in rapidly growing economies. This focused on the need for a sustainable set of relationships between water, power and agriculture ("the nexus").

> Business leaders at the World Economic Forum Annual Meeting in 2008 set out a Call to Action on Water, to raise awareness and develop a better understanding of how water is linked to economic growth across a nexus of issues and to make clear the water security challenge we face if a business as usual approach to water management is maintained. This report captures where the debate is now and sets out the challenge we face if nothing is done to improve water management in the next two decades (World Economic Forum 2014).

While the practical mechanisms to address these newly defined challenges remained unclear, the emergence of the "nexus" concept offers the opportunity to reconnect water resource planning with broader development planning; although it focuses primarily on the agriculture and energy sectors, practical approaches may spill over into other sectors. Unfortunately, nexus thinking has subsequently been driven primarily by the water sector, with little or no input from the energy and agricultural sectors, thus negating much of the potential of the nexus to reconnect water to broader development planning in a hydro-supportive manner.

More recently there has been a shift in African perspectives on infrastructure investment, driven by the demand from African Ministers, through their African Ministers Council on Water (AMCOW), for water to contribute more to the continent's growth and development as well as by the recognition that Africa is the 'under dammed' continent (African Development Bank).[1] From an African perspective, perhaps the most important development has been the emergence of new sources of finance for large water infrastructure projects from China, Brazil and India which, more than any other intervention, has changed the water resource management discourse. Many donor-dependent countries now have alternative sources of assistance. They need no longer spend years in stakeholder consultations to justify clear infrastructure requirements and can often get responses that, while not always positive, are rapid in comparison to their experience with their traditional western development agencies. As a consequence, the rate of investment in large water infrastructure has increased significantly, the "revealed preferences" providing evidence of the impact of prior investment boycotts.

The contribution of China to this changing dynamic has in turn seen international environmental organisations, make considerable efforts to influence China's policies. The strategy of international NGOs to link with business and finance partners to influence a major government is an interesting innovation in the broad strategy of promoting global regulatory harmonization which has been described by Drezner (2007).

---

[1]Despite having two of the largest dams in the world, SADC States have an average per capita storage of just of 500 m^3/person, against the global average of 1500 m^3/person.

3 Water in Mainstream Development Planning—Could Virtual Water and Nexus Contribute?[2,3]

As with water resource planning, national development planning has had a che-quered history. In the 1960s, it was the mainstream approach in many countries, particularly newly independent developing countries but also in a number of developed economies. It declined in importance, in part for ideological reasons. But:

> There were also well-founded concerns about the performance of planning since the out-comes often fell far short of the objectives. There was a variety of reasons for this, from unrealistic assumptions about internal capabilities and external markets as well as slow responses to external pressures such as the oil price shocks of the 70s. These problems were compounded in many cases by weak governments that were unable to link planning theory to implementation practice while economic technocrats, often from abroad, dictated development paths with little attention to local social and political geography (DBSA 2012).
>
> […] However, the legitimacy of the idea of planning for development was sustained by the fact that the countries that proved best able to navigate the global financial turmoil of the 1990s turned out to be the East Asian "tiger economies" whose centralized planning systems were an important contributor to their economic success (DBSA 2012).

There has been a revival in planning for development but in a modified form. Poverty Reduction Strategy Papers (PRSPs) addressed not just the socio-political impact of structural adjustment programmes but also helped to re-establish a budget framework and development strategy for donor-dependent countries. These were, however, short term measures:

> There was a clear need in many countries for a better structured, more generic, long term development framework and the institutional arrangements to prepare and maintain it. Indeed, it has been argued that few developing countries have made significant economic progress without a long term development plan. A more substantive set of approaches has emerged which seeks to frame longer term and more comprehensive development pro-grammes. They continue the trend away from detailed long term forecasting and avoid engaging in the detailed decisions on individual projects and investment allocation and focus rather on countries' strategic direction (DBSA 2012).
>
> These approaches go beyond technocratic efforts to identify global and national trends, to identify interventions and allocate resources to take advantage of them. Rather, they recognize the need, in complex societies, to bring focus to and generate consensus around key national priorities and coherence in pursuing them, mobilizing support from broad sections of society rather than simply managing governmental action. To the extent that they address development strategy their focus is on the development of long-term national visions and then seeking strategies to achieve them, built on an understanding of local endowments, challenges and opportunities (DBSA 2012).

---

[2]This section is drawn from the discussion document for a workshop for national planning agencies of SADC countries on Understanding National Development Planning and its Contribution to Inter-Sectoral Regional Integration, organized by the NPC and DBSA in August 2012.

[3]DBSA (2012).

In this new approach, the plan is a process rather than a product; it is effective to the extent that there is political leadership in its development, substantive involvement of the institutions concerned in its elaboration and discipline in its implementation.

> [...] the plan can only be as good as the quality of the policies that are in it, which in turn will be largely determined by the quality of the institutions, in government and beyond, that contribute to it [...] A useful contribution of development planning has been to force sectoral agencies to consider the feasibility of their policies and proposals in the broader national context. In this sense, development planning can contribute to institutional strengthening (DBSA 2012).

A critical feature of national development planning in SADC is that it is conducted within a country political framework where national governments have direct authority over the public sector and considerable indirect suasion over other stakeholders since they set the direction of both regulatory and public spending interventions.

The political environment for planning at regional level in Southern Africa differs from that at national level primarily because it is undertaken on a cooperative basis without the benefit of direct political authority and with no system to hold national governments to account if they fail to meet their obligations. Inter-sectoral coordination is a particular challenge. While national development planning, which falls under the authority of a head of state and single executive, can achieve integration between sectors at national level the same is not true at a regional level. While decisions may be taken and announced, implementation may falter if regional discussions and decisions have not been adequately informed by national considerations. For this reason, regional plans are often not acted upon, as is highlighted in the electricity sector in SADC and described in the Chapter of this book on "Electrical Power Planning in SADC and the Role of the Southern African Power Pool". This has been identified as a generic underlying issue by SADC in the course of its review of the progress made with its 2005–2015 RISDP.

To date, SADC's main successes have occurred where cooperation has been required between single, inter-linked sectors. So transport networks, electricity grids and telecommunications systems have evolved with some degree of success. This reflects the abilities of single sectors to convene to identify areas of mutual interest and cooperate to address them, a process which regional agencies such as SADC can facilitate.

In areas where inter-sectoral cooperation is required, progress has been notably slower, perhaps because of the higher transactional costs, but also perhaps a result of a lack of a clear regional framework for cooperation for mutual benefit. So cooperation in agricultural development, which requires transport, trade and, potentially, water sector support has been less successful, judging by trade flows. The extent to which national interests may conflict in cross-sectoral planning is also greater; for this reason, progress in trade in services has also been slow.

Regional development planning is thus usually of a consultative and indicative nature. In this context, sectoral planning will still reflect national priorities and trade-offs between sectors, while, in strategic sectors like water, trade and energy, efforts to promote *regional best options* will have to address sovereign security concerns as well as to manage the influence of national interest groups.

National development planning brings together the different sectors within an overall framework of policy and strategy and seeks to identify and address potential linkages, synergies and constraints between them as well as to make trade-offs between different priorities. A critical question is the extent to which regional cooperation and integration are included as objectives in national development planning processes and efforts made to ensure that development strategies are coordinated. A formal process of coordination would help to identify costs and benefits of regional policies at national level and guide negotiations and decision making. A review of national development plans in SADC found significant variation; while some national plans had entire chapters on regional integration, others ignored the subject completely. It has been suggested that approaches that could more effectively mobilize national development planning in support of regional integration need to be developed and implemented. The potential advantages of considering regional best opinions have rarely been effectively quantified and there has been little effort to address sovereign security concerns which are clearly justified, given ongoing instability in some SADC countries.

4   Water in Development Planning, National and Regional

All Southern African countries' national development plans address water and related issues and many make clear linkages between water and energy and water and agriculture, although not always in a coherent manner. The expansion of irrigation has long been recognised as an important intervention to increase the productivity and reliability of agriculture. The practical examples of Kariba and Cahora Bassa hydropower installations have highlighted the potential of water to produce energy although, aside from flood control, not the potential multi-purpose opportunities.

Beyond the Zambezi dams, South Africa has long dealt explicitly with the water-energy nexus, the strategic outlines of which were spelt out in the 1970 Commission of Enquiry report on water matters; interestingly, that report was not unduly concerned with the potential impact of water scarcity on irrigated agriculture, concluding simply that increased water use efficiency in agriculture would address most of the growing pressures. More recently, the national Department of Water Affairs identified the potential role of regional cooperation in agriculture as a strategy to address water constraints (see Box 2). However, the economic evidence is that South Africa will be a net exporter of agricultural products (some of which will be irrigated) for decades to come.

In other Southern African countries, there is an understandable priority for water supply and sanitation matters although there is increased emphasis on hydropower, both as a consequence of the failure of regional cooperation to provide energy security as well as of the success of efforts to develop mining. Given the donor emphasis, language on IWRM is also prevalent in the water chapters of national development plans—one consequence of this is that much water-related development is addressed in the planning of other sectors, notably power and agriculture rather than by water authorities.

Text box 2 Practical Approaches to Regional Water-Food Issues (see Footnote 1)

Practical approaches to regional water-food issues

South Africa uses 60 % of its scarce water resources on irrigation, a substantial portion of which is used to irrigate crops which are regarded internationally as rain-fed crops. The question is therefore being asked about the extent of alternative production areas in southern Africa (particularly in selected neighbouring countries) for the range of crops which are presently produced sub-optimally under irrigation in South Africa. The objective of this study is therefore to provide an answer to this question with adequate confidence to allow the rational pursuit of this concept which could have far-reaching mutual benefit for southern African countries. The countries that were considered are Mozambique, Zimbabwe, Malawi and Zambia.

This broad assessment revealed that the four target countries possess a net area of about 26.6 million ha of high-potential rain-fed cropping land (referred to as "Premium" land use potential) with the following breakdown per country: Zambia 11.1 million ha; Mozambique 8.8 million ha; Zimbabwe 6.3 million ha; Malawi 0.4 million ha. The constraints include land tenure issues (the majority of the high potential rain-fed cropping area is occupied by subsistence farmers on communally owned land), population (the high rural population spread presents a challenge to commercialisation of agriculture), present land use (widespread subsistence farming), poor or lacking infrastructure and poor agricultural support services. However, the constraints are not considered insurmountable. With the appropriate vision, investment and support from the governments of the respective countries there are significant opportunities for extensive commercial agricultural development which could involve and benefit local farmers and their communities. The recent examples of South African farmers operating successfully in Mozambique and Zambia, with full government backing, have shown that these constraints can be overcome.

Whilst the principal objective of this study is to identify areas that are suited to rain-fed crop production, the existence of a considerable network of largely "un-tapped" surface water resources, especially in Zambia and Mozambique is highlighted. There is therefore an opportunity for expanded utilisation of the water resources in these countries for irrigation where there is a higher irrigation potential, in terms of both soils and climate, than exists for many of the irrigation areas of South Africa.

(Ex: DWA (2010))

The current SADC focus is water-centric, reflecting SADC's overall approach. Thus it has promoted the establishment of river basin organisations and encouraged them to engage in sector-led, basin-bounded planning exercises. Beyond contributing to a better understanding by water practitioners of their water resources, this focus has not helped national water sector agencies to engage with their own national development processes nor undertaken work at the regional level that could support that kind of endeavour.

This approach reflects both donor preferences (strongly expressed by the provision of technical assistance under the control of donor officials) as well as SADC's generic working models. However, it is becoming clear that these approaches are not producing significant results.

Major projects are proceeding (or stalling) without significant contribution from the regional water sector. Zambia is developing its hydropower resource on a national (or, in the case of Kariba, bilateral) basis and waited until most of the projects were underway before ratifying the Zambezi Watercourse Agreement in 2013. Development of the Batoka Gorge and Mphanda Nkuwa projects on the Zambezi is also being led by the power sector on a bilateral basis, with only limited input from a water resource management perspective.

Recently (2013), SADC convened an investment conference for the water sector which was poorly attended, not least because the major projects presented were already well known and under development through other channels while smaller projects appeared to reflect national wish-lists rather than strategic projects of regional significance.

The challenge for hydro-centric processes is to convene not just water sector representatives but also stakeholders from other sectors. Globally, few regional water institutions have any sovereign authority either to convene or to take decisions in respect of water management and use. The exceptions are the European Union which has an overarching political framework and the Senegal River basin where governments have formally delegated specific water management powers and responsibilities to a joint water management institution.

Even if there were substantive political framework, it would only be effective if the regional representatives of the different sectors were adequately briefed on the national issues and inter-sectoral trade-offs. In the absence of such a framework, it is necessary to place greater focus on generating and sharing information and participating in other sectors' processes and less on trying to tell other sectors what to do and how to organise themselves.

Hydro-supported planning in water resources focuses on identifying and engaging with strategy and planning activities in key user sectors. Where this has occurred, there have been notable successes. One example is the Lesotho Highlands Water Project, which emerged from engagement with urban and industrial users, during which it became clear that the demand for water would increase beyond the ability of the Vaal system to support it.

At a smaller scale, Swaziland's agricultural development required additional water to enable its sustainable expansion; the LUSIP project became one of the catalysts that led the national water sector institutions to negotiate the Interim IncoMaputo Agreement which was signed in 2002.

After many years of argument, Namibia has now indicated that it intends to proceed with plans to tap the Okavango river to meet its development needs, despite continuing objections from environmental interests.

In the agricultural sector, there is renewed interest in water as a factor of production that has potentially opened the way for greater collaboration with water resource managers. There is however as yet little evidence to suggest that this is being translated into practical action. Similarly, while multi-sector modelling has demonstrated the potential of synergies on the Zambezi river between power, agriculture and environmental conservation, this has still to be translated into terms which the user sectors relate to—for example, in the power sector, there is a concerted move towards ensuring energy self-sufficiency even as the water-related studies demonstrate the benefits to be reaped from cooperative development and management.

One reason for the failure to make more progress with regional cooperation and integration in the water sector is the institutional and transactional demands that it imposes. This is a generic challenge. Integration cannot simply be driven by a single regional organisation. Many of its elements have to be implemented cooperatively by sovereign national governments. If its potential benefits are not understood—and preferably experienced in a practical way—by a significant proportion of a country's citizens, it will be hard to convince them to support it.

Judging by the slow progress made to date, Southern Africa's regional and national institutions have not generally succeeded in demonstrating those potential benefits. The problems with SADC's approach are recognised by the organisation itself and are generic and not limited to water. The organisation's own recent assessment includes, amongst ten "lessons learned" that:

> There is no effective link between the SADC Secretariat, the SADC National Committees and relevant key stakeholders who are supposed to oversee and effectively implement SADC activities and programmes at national level (SADC 2011).

A failure to engage with broader development priorities and to focus instead on water centric issues has been blamed for the failure of the approach, most recently in the Mekong river basin where coordination efforts have been ongoing for over 50 years. As the former CEO (2004–2007) of the Mekong River Commission (MRC) has commented,

> Hydro-diplomacy tends to be more environmentally than economically oriented [...] since the signing of the "Mekong Agreement" in 1995, donors have oriented MRC's activities mainly toward information and knowledge management, while downplaying its investment facilitation role.
> With such a vision of the role of basin organizations, there is a risk that they will continue to be excluded from the national investment planning process. Governments will continue to complain about the lack of tangible results for the direct benefit of the population. They will also remain reluctant to increase their financial contributions.
> Basin organizations may well get stuck [...] playing an insignificant role in the negotiations about the most critical issues. No doubt that knowledge is essential for informed

decision-making, but its generation and communication should first and above all be developed at national level, on the basis of the subsidiarity principle (Cogel 2014).

This in spite of the fact that the four countries of the Mekong River Commission signed an 'Agreement on Cooperation for the *Sustainable Development* of the Mekong River Basin' [emphasis added] and Article 2 of that agreement calls for "with emphasis and preference on joint and/or basin-wide development projects and basin programs." Indeed the first prior consultation process under the 1995 Mekong Agreement, the Xayaburi Hydroelectric Project, focussed on hydro-centric concerns about potential impacts on the mainstream of the Mekong, and not on the contribution to regional energy security and growth. This process failed to establish any clear agreement on the acceptability of the project on that basis.

5   Political Economy of Regional Development Planning

Energy: As outlined above, the determination and evaluation of the opportunities and constraints posed by water resources—and other natural resource endowments—involves coordination between different political jurisdictions and across multiple sectors whose priorities and criteria, implicit and explicit, may be expressed in a range of different metrics.

While at a national level, development planning processes can establish a common metric to assess costs and benefits, this is more difficult to do regionally, across a diverse set of administrative systems. So while apparent benefits that could be achieved through regional planning and cooperation have often been identified at a conceptual level, it has proved difficult to detail their practical implications at a national level. As a consequence, many apparent opportunities have not been acted upon.

An example is provided by the power sector. According to the economic metric, the region would benefit considerably (in terms of cheaper energy) if a regional perspective was taken and a complementary suite of generation projects promoted (see Chapter "Electrical Power Planning in SADC and the Role of the Southern African Power Pool" for the details). In practice however, this has not occurred. Aside from the economic analysis of investments and operating costs, other metrics have been introduced. So countries are concerned about reliability of supply and their experience has been that there are higher risks to dependence on neighbouring countries than on their own capacity.

This situation has led to a preference for sovereign (national) rather than regional solutions—and indeed, a rejection of proposals for greater cooperation, despite the apparent benefits that they offer. In this case, a second-best regional strategy has emerged from CRIDF—once all countries have adequate generating capacity to meet their needs, they may use the regional power pool to trade and to purchase cheaper electricity if it is available elsewhere. This may realise financial gains for the sellers and buyers, reduce regional carbon emissions, and realise some modest water savings, an illustration if not a product of the nexus and Virtual Water approach, with power not water as the driver. But it will be based on a sub-optimal investment strategy which has built more capacity than needed.

Agriculture: There is already extensive recognition of the potential for regional synergies in agriculture and for water to be exploited to strengthen regional food security at country level (see Box 3). However, if there is to be support for exploiting the extensive land, water and human resources outside of South Africa to produce food for the region, local metrics will have to guide the argumentation and prioritisation.

In most cases, a priority will be to ensure that agricultural development is accompanied by livelihood enhancement—certainly that livelihoods of poor rural populations should not be undermined. To the extent that the resource outside of South Africa is developed using farming models that expand livelihood opportunities for small scale farmers, this should also contribute to household level food security.

In this context, any support by CRIDF to the development of resilient, more productive, small scale agricultural production in the region will enhance resilience and food security across the region as well as providing direct household benefits. The regional benefit of these approaches will depend on wide-scale replication, whose local impacts and cumulative effects will have to be carefully assessed. While it may be possible to describe this in terms of Virtual Water and the nexus, and investments in water infrastructure may be a necessary part of such a strategy, they will only be complementary to the wider challenge of the establishment of farmers with the appropriate skills as well as the development of the farming systems, markets and support institutions and enabling infrastructure required to enable competitive production and trade to occur.

Similarly, mobilising the benefits of locating agriculture to take advantage of higher rainfall, and hence reduce the dependence on blue water will also require significant investment in other (non-water) infrastructure and institutions. Virtual Water and nexus thinking may help to highlight the need for hydro-supportive integrated national planning into perspective, and may introduce other options and trade-offs to this process.

Text Box 3 Trade-Based Food Security in South Africa's National Development Plan

Trade-based food security in South Africa's National Development Plan

"It is necessary to make a distinction in policy discourse between "national food self-sufficiency", "food security" and "access to food by poor people". South Africa is food-secure and has been for a number of decades. This means that it earns a trade surplus from agricultural exports and is able to cover the cost of food imports from those exports. The country has also produced enough of the staple cereal (maize) for all but three of the past 50 years (the exceptions being the droughts of 1984, 1992 and 2007). The composition of the maize harvest is changing, however, with more yellow than white maize planted. This reflects the trend towards higher consumption of animal proteins and the fact that wheat, rice and potatoes are becoming the preferred staples as the population urbanises and becomes more affluent. In this regard, the national food-security goal should be to maintain a positive trade balance for primary and processed agricultural products, and not to achieve food self-sufficiency in staple foods at all costs.

Region-based approaches to food security should be investigated. As South Africa's agriculture becomes more specialised and efficient, there may be a trend away from the production of staples to higher-value crops. As there is only limited correlation between climatic events in South Africa and countries to the north of the Zambezi (although the drought of 1991/92 was regional in nature), regional cooperation may offer greater supply stability and resilience to droughts. Regional economic integration is best served when there are complementary interests and advantages between the parties, which may be the case in food production. Regional expansion of production, as seen in recent years, is favourable. South Africa should benefit from the opportunities this brings for trade, food stability and value-chain consolidation." (National Development Plan: p. 230)

These examples highlight the general principle that successful regional cooperation and integration depends on a clear identification and equitable and reliable distribution of the costs and benefits of any regional development initiative. One advantage that has been posited for "top-down" institutional structure of regional integration rather than ad hoc sectoral "bottom-up" approaches is that it is easier to negotiate packages of initiatives with an acceptable mix of costs and benefits in a multi-sectoral context than in a single sector. The high level of coordination that this requires both within regional institutions and between national and regional institutional families continues to present a strategic challenge to the achievement of the broader regional integration goal.

6 Conclusions

It has already been demonstrated that mobilising synergies and exploiting complementary resource endowments between countries could increase the productivity of agriculture and power production and reduce risks due to climate variability and change, potentially benefitting a range of economic interests and communities across the southern African region. The major challenge remains to give effect to this approach.

Some of the policy synergies would reflect the concept of Virtual Water by encouraging agricultural production in most favoured and least vulnerable areas. The nexus could be reflected in increased availability of relatively reliable and "green" hydropower to countries of the region, traded through the Southern African Power Pool (SAPP). Similarly, trade-offs between irrigation and hydropower, albeit on a temporary basis during drought, could be informed by nexus and Virtual Water thinking. This should make it possible to enhance both food security and energy security for poor people in the region although that outcome would not necessarily be automatic.

However, decisions about the adoption of such policies will be taken primarily at national level and will depend on the political economy in each country. While regional cooperation may play a role and can certainly inform the process, the costs, benefits and trade-offs will need to be acceptable at each level of decision-making.

The implications for policy advocates is that, while regional institutions may be useful to develop understanding of potential synergies and channels through which to communicate this information, greater attention should be paid to national political economies and to national costs, benefits and trade-offs.

Development planning processes could make an important contribution to elaborating such multi-component regional integration "packages" but are still in their infancy in SADC and structures and methodologies that allow the various inter-sectoral trade-offs at national level to inform decision-making about regional integration have yet to be established. This imposes constraints on the potential for the development of cooperation on water-related opportunities.

In this context, the use of hydro-centric approaches to water resource planning within shared river basins rather than encouraging cooperation at the level of national economies may weaken cooperative inter-sectoral work since it tends to place water above and apart from mainstream planning processes.

The political economy that determines whether potential economic and social benefits are translated into political decisions remains poorly understood although it has been identified as a priority area for further research.

In this broad context, the concepts of Virtual Water and the nexus may usefully inform a range of discussions and be used to illustrate potential challenges of and responses to climate change. They are however unlikely in themselves to provide the basis for national policies on which regional cooperation and action depend, given the many other factors that have to be considered.

References

Biswas C (2012) Preface. In: Tortajada C, Biswas C, Altinbilek D (eds) Impacts of large dams: a global assessment. Springer, Berlin
Cogels O (2014) Hydro-diplomacy: putting cooperative investment at the heart of transboundary water negotiations. In Pangare G (ed) Hydro diplomacy. Sharing water across borders. Academic Foundation, New Delhi
DBSA/NPC (2012) Understanding National Development Planning and its contribution to inter-sectoral regional integration (Discussion document for workshop of SADC country national planning agencies held at DBSA in August 2012). DBSA/National Planning Commission

Drezner DW (2007) All politics is global: explaining international regulatory regimes. Princeton University Press, Princeton

DWA (2010) An assessment of rain-fed crop production potential in South Africa's neighbouring countries. 2010 report: P RSA 000/00/12510. DWA, Pretoria

European Union (2000) Water framework directive. Online at http://www.europa.eu.int/eurlex

Global Water Partnership (GWP) (2000) Integrated water resources management. TAC background paper no 4. Global Water Partnership, Stockholm

Grey D, Sadoff CW (2007) Sink or swim? Water security for growth and development. Water Policy 9(6):545–571

ICWE (1992) The Dublin statement on water and sustainable development. In: International conference on water and the environment (ICWE), Dublin

Manuel T (2011) Keynote address at the workshop on "Regional approaches to food and water security in the face of climate challenges", 23–24 May. DBSA, Midrand. Online: http://www.npconline.co.za/pebble.asp?relid=590

Muller M (2015) The 'nexus' as one step on the road to a more coherent water resource management paradigm. Water Alternatives 8(1)

SADC (2011) Desk assessment of the RISDP 2005–2010, Gaborone

Sadoff C, Muller M (2008) Water management, water security and climate change adaptation: early impacts and essential responses, (Background paper No. 14) GWP, Stockholm, Sweden

UN (United Nations) (1977) Report of the United Nations water conference, New York, 1977. The document is not available on the UN website but can be found at http://www.ircwash.org/sites/default/files/71UN77-161.6.pdf. Accessed May 2014

World Economic Forum (2014) Water security: the water-energy-food-climate Nexus. Accessed at http://www.weforum.org/reports/water-security-water-energy-food-climate-nexus

Mechanisms to Influence Water Allocations on a Regional or National Basis

Barbara Schreiner

Abstract One of the more pervasive challenges facing SADC is the variability in water availability over space and time. In response, SADC States have largely opted for large scale water storage and transfer infrastructure. However as climate and demographic changes, together with rapid economic growth, increase water stress and variability, larger scale infrastructure solutions will become increasingly expensive and environmentally unsustainable. As more basins face closure this will become a regional rather than national challenge, raising pressures around the reasonable and equitable use of water. A Virtual Water and nexus perspective may offer a different view for national planners, making better use of the total water footprint to support economic growth, as well as to meet social and environmental needs. This will require the allocation (or re-allocation) of water between sectors to formalise and regulate entitlements which optimise allocations across the full water footprint and which balance the water requirements of food and energy production. Similarly, inter-state allocations based on the reasonable and equitable use principles espoused in the SADC's Revised Protocol on Shared Watercourses, may have to consider both the blue and green water contributions to the national economies of the riparian states. On a regional basis, while there has been some attention paid to importing of Virtual Water in agricultural products rather than to produce them locally, this concept has not gained much traction. It is therefore perhaps unrealistic to suggest that Virtual Water and Nexus based allocations per se would be a viable option to introducing the concepts to SADC. Nonetheless, the author argues that the introduction of the concepts into the national planning processes places other options on the table, promoting better trade-offs and an improved understanding of the contribution of water to the economy as a whole.

B. Schreiner (✉)
Pegasys Institute, Cape Town, South Africa
e-mail: barbara@pegasysinstitute.co.za

© Springer International Publishing Switzerland 2016
A. Entholzner and C. Reeve (eds.), *Building Climate Resilience through Virtual Water and Nexus Thinking in the Southern African Development Community*, Springer Water, DOI 10.1007/978-3-319-28464-4_5

1   Introduction

One of the more pervasive challenges facing the SADC region and its member states is the variability in water availability across the region and over time. To date the adage, "water allocations follow the economy" (or hydro-supportive approaches as referred to by Muller in Chapter "Virtual Water and the Nexus in National Development Planning") seems to have held in responding to this challenge. Generally engineering solutions have been developed to get the water to where the formal economy needs it, as well as to provide for basic water supply and sanitation services. In this scenario, factors other than water costs drive both the direction and location of economic growth, while water managers respond to these needs through infrastructure solutions.

However, as climate and demographic change, together with economic growth, increase water stress—larger scale engineering solutions will be demanded. Water delivered through these schemes will become increasingly expensive and environmentally unsustainable. As more basins face closure, and as variable water supply becomes a regional rather than national challenge (see Chapter "The Future of SADC: An Investigation into the Non-political Drivers of Change and Regional Integration"), larger international inter-basin transfers may be seen as the solution to regional water challenges. Turton's analysis in Chapter "The Future of SADC: An Investigation into the Non-political Drivers of Change and Regional Integration" highlights some of the planned schemes (particularly transfers from the Zambezi Basin) in this regard. These schemes appear to be underpinned by the notion that SADC's wetter northern countries have more than sufficient water, and that royalty systems (like that between Lesotho and South Africa) hold net benefits for all parties. Conversely, however, the World Bank's Multi-Sector Investment Opportunities Analysis for the Zambezi basin shows that increased abstraction from the Zambezi can have significant impacts on the hydro-power potential of the basin (World Bank 2010). Nonetheless, a stakeholder to the development of the Integrated Water Resource Plan (IWRM) for the Zambezi noted that "*Consider taking a regional perspective rather than a Basin perspective in the analysis of projected water use, as there's a strong demand for Inter-Basin Water Transfers in Southern Africa.*" (Mott-Macdonald 2008).

However, the transboundary implications of these schemes may raise regional pressures around the reasonable and equitable use of water, particularly in times of drought. The better watered northern countries are likely to want to retain access to water for future use. As a regional expert noted in the Zambezi IWRM Strategy "*.... whereas actual water use in the basin is presently small, the actual demand is much higher. We should not assume there is plenty of water to go around.*" (Mott-Macdonald 2008). Ultimately, the drier southern nations may have to push water towards sectors that provide more income and jobs per drop, rather than rely on large regional transfer schemes to sustain economic growth. The management and allocation of water both nationally and regionally will therefore become increasingly important.

Hard choices will consequently have to be made between allocations to various water use sectors and between sovereign states. The hypothesis in this regard is that a Virtual Water perspective may offer a different view for consideration by national planners, providing an alternative to large scale regional water transfer infrastructure and making better use of the total water available to support economic growth and rising water demands. These approaches may help build regional resilience to variability in water availability and climate change, and provide the necessary food, energy and environmental services across the region. Ultimately, an argument will be made that this may further strengthen cooperation, avoiding the potential regional tensions that may lie in large scale north to south inter-basin transfers.

The allocation of water between and within states and sectors will play an important role in effecting this scenario. Virtual Water thinking recognises the blue, green and grey water components of water use and their implications in decisions on water allocation. The cross-cutting themes of the availability of allocable water and the requirements for assured water supply to encourage and sustain investment, add complicating dimensions.

This paper outlines the mechanisms and opportunities to shift water use patterns on a national and international basis in the SADC region, set against the broader water challenges for the region, to respond to potential benefits that may be gained through the Virtual Water and water, food and energy nexus perspective. It addresses the challenges facing these allocation and re-allocation processes both with respect to allocations between sovereign states as guided by the Revised SADC Protocol and basin specific agreements, and allocations between users and sectors as provided for in national legislation.

2 Current Situation

There is significant hydrological variability in the SADC region from the humid well-watered Democratic Republic of the Congo (DRC) to the arid regions of Namibia and Botswana. In addition to the spatial variation, there is significant temporal variation over the region as well. Figure 1 shows the extreme vulnerability of the region to droughts, while Fig. 2 shows the vulnerability of the eastern part of the region in particular to floods.

In addition, Fig. 3 shows the levels of economic and physical water scarcity in the region. A significant portion of the SADC region suffers from economic water scarcity, while most of the rest of the region is approaching or experiencing physical water scarcity. The management of water in the region is thus critical.

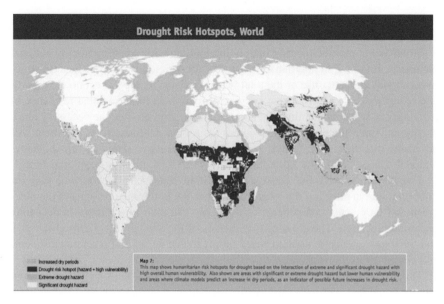

Fig. 1 Drought hotspots (*Source* Care Climate Change 2015)

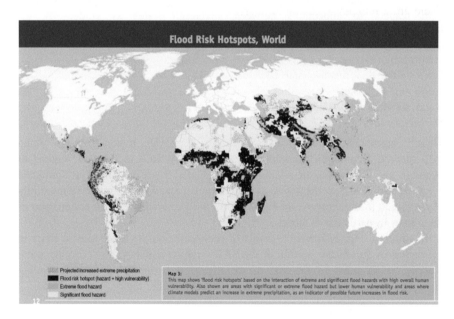

Fig. 2 Flood hotspots (*Source* Care Climate Change 2015)

Global physical and economic water scarcity

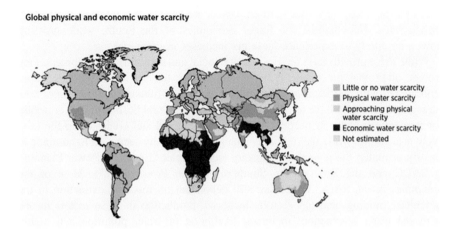

Fig. 3 Areas of physical and economic water scarcity

3 Drivers of Change

There are a number of drivers of change in the SADC region which are placing increasing demands on the water resources of the region. Several countries are experiencing unprecedented GDP growth rates in primary economic sectors (agriculture, mining, oil and gas), and the International Monetary Fund (IMF) places three of SADC's member states in the top ten fastest growing economies of the world. This will have a significant impact on the region's water resources for both water demand and quality. While GDP growth rates are variable throughout the region, Angola, the DRC, Mozambique, Tanzania and Zambia are experiencing growth rates of over 6.8 % (see Fig. 4). A further complicating dimension therefore

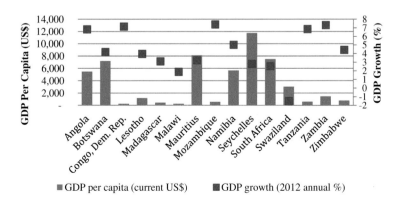

Fig. 4 Per capita GDP (US$) and GDP growth (2012 annual %) for SADC countries

lies in the fact that the wetter northern basins are largely experiencing these rapid growth rates. Nonetheless, the larger economies of the South, while growing slower, are likely to experience large net increases in water use.

While these growth rates are positive for social and economic development, they provide additional pressures on water and energy (electricity) resources. Rapid economic growth, particularly in the extractive sectors, fundamentally changes how the region is going to have to address water issues. Several basins that are currently not stressed are likely to become more heavily utilized and contested in the relatively near future, while the regional demand for electricity seems set to continue to outstrip or match the generation capacity (see Chapter "Electrical Power Planning in SADC and the Role of the Southern African Power Pool"). Most of the economies in the SADC region are still dependent on resource extraction in the agriculture, mining, and gas sectors. Increased production in these sectors means increased water abstraction, increased likelihood of water pollution and higher energy requirements with associated water requirements.

Hydropower is already major source of energy in the northern SADC countries, and there is significant potential for further hydropower potential. The current installed hydropower capacity in the Zambezi is 5000 MW, with the potential for an additional 13,000 MW. Evaporation off the existing hydropower reservoirs is seen (in the Zambezi IWRM plan) as the single largest user of water and the Zambezi River Authority allocates water volumes to the generation facilities in Zambia and Zimbabwe. However, this hydropower comes at considerable climate risk as increased evaporation from higher temperatures and reduced runoff limits the hydropower capacity. Plans for an increase of some 184 % in the area under irrigation in the Zambezi add a complicating dimension for allocations between hydropower and irrigation, with transboundary implications (World Bank 2010). Zambia's energy policy is nevertheless set to become a net energy exporter through hydropower, while its national growth and agricultural strategies aim to increase the contribution of irrigated agriculture to GDP. Zambia, as one of the key countries of the Zambezi basin, generating up to 40 % of the runoff, is also virtually entirely dependent on hydropower as a source of electricity.

Climate change also has major implications for water management elsewhere in SADC. The hydrological cycle is the primary pathway through which climate change impacts are felt. The basins of the region are expected to experience more extreme floods and droughts as the climate changes, as well as significant increases in temperature. While, most global climate models indicate increases in temperature in the inland areas of SADC, the impacts of this on rainfall and runoff are much more tenuously understood, and appear to depend on the position of the Inter-Tropical Conversion Zone (ITCZ), which lies across Angola, Zambia, Malawi and Tanzania.

Population growth and migration patterns are also placing changing demands on water resources. The proportion of the SADC population that is urban has changed from 33 to 40 % over the last decade, while the total population has increased from some 220 million in 2001, to 281 million in 2011 (SADC 2011)—an increase of some 20 %. In the year 2000, freshwater withdrawals in SADC mainland states (excluding Zimbabwe) totalled 18.78 billion m^3, in 2009 withdrawals totalled 20 billion for these same states (SADC 2011)—an increase of some 10 %. Over 50 % of the total withdrawals are in South Africa, but South Africa holds only some 1.4 % of the total renewable water resources of the SADC mainland states, with around 21 % of the population.

These changes are taking place in a context of planned regional integration in SADC and Africa more broadly. This has a direct impact on water resources of the region, both with respect to shifting the main regional water demand centres as well as changing water availability unequally across the region. Optimal regional use of water in support of the integration agenda has considerable implications for the allocation of water to the agricultural, energy and mineral extraction sectors, but also provides strategic opportunities when taking a Virtual Water perspective. The import and export of goods that require significant water in their production may prove an alternative to the typical inter and intra-state water allocation processes.

4 Water Allocation Mechanisms

Water allocation is the process of sharing this limited natural resource amongst competing users, sectors or administrative regions. Allocation becomes increasingly necessary when the natural distribution and availability of water fails to meet the needs of all water users—in terms of quantity, quality, timing of availability, or reliability. In simple terms, it is the mechanism for determining who can withdraw water, how much they can take, from which locations, when, and for what purpose (Speed et al. 2013).

There are different allocation mechanisms that operate at the inter- and intra-state level. The former, in the SADC context, is governed by the revised SADC Protocol on Shared Watercourses, or basin specific arrangements that specify required cross border flows such as the Interim Inco-Maputo Agreement, or the Lesotho Highlands Water Treaty. The latter is governed by the specific allocation mechanisms established by national policy and legislation. These give rise not only to sectoral allocations, but also to water use authorisations for specific users. Figure 5 shows the scaling of these forms of allocation.

At the basin level in SADC, water allocation processes between riparian states are covered by Articles in the Revised Protocol requiring the notification and reasonable and equitable use of water, as well as the permitting of waste discharges. In some cases, for example the Interim Inco-Maputo Accord and the Lesotho Highlands Treaties, international treaty governs actual cross border flows. National water allocation processes are governed by various national water laws.

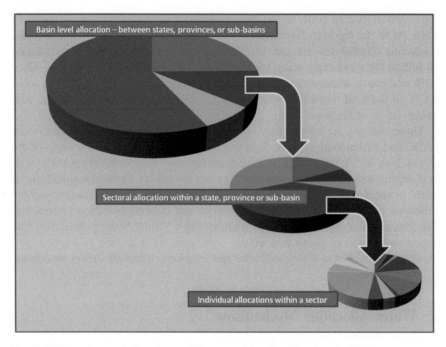

Fig. 5 Different forms of allocation at different spatial scales (Speed et al. 2013)

There are various different approaches to the allocation of water including:

- *"Automatic entitlement"*. Some water allocation processes recognise an automatic minimum entitlement to water for basic social purposes, or the maintenance of minimum environmental requirements. This is noted in most SADC mainland state water legislation at varying levels of sophistication.
- *"Administrative or bureaucratic process"*. The right to abstract water is given by some authority, either a state agency or a user group (e.g. an irrigation board). This is the most widespread formal type of allocation process.
- *"Communal or traditional processes"*. An enormous range of allocation process exists that are based on traditional, non-state law or custom.
- *"Market allocation"*. In some parts of the world, water rights are reallocated on the basis of trade rather than by administrative allocation. Both formal and informal water markets exist.
- *"With land"*. Water rights may be attached to the ownership of land. Transfer of the land through sale or inheritance implies transfer of the water right. In some cases, landowners abutting a surface water resource are entitled to water rights. Similarly, groundwater below private property is often regarded as an entitlement of that property (Speed et al. 2013).

In the SADC context, since the wave of water reform in the 1990s, formal water allocation systems are largely based on administrative systems, through licences or permits, with automatic entitlement for subsistence level activities. However, in large parts of the region, communal or traditional water allocation processes also still operate, largely outside the formal legislative systems.

South Africa has allowed a system of trading of water use entitlements for many years, although this is now under threat from a policy proposal to removal the option of water trading from the legislation. In Zambia, water trading is not recognised under the water policy, but is encouraged under the agricultural policy.

No single means of allocating water takes preference over another and a plurality of water rights systems operate in any basin, starting from the transboundary level and cascading through the national, often to local systems. Although in most countries in SADC, basic human needs and environmental water are recognised as priority allocations. The suitability of an allocation system is determined by local conditions, including the capacity of the state to effectively administer the system. In this regard, "any effective allocation mechanism is entirely dependent upon the development of significant institutional capacity from the national to the catchment level ... to assess available resources and any necessary ecological requirements, and administer, monitor and enforce the water allocation process" (Speed et al. 2013).

In addition to institutional capacity, the effectiveness of allocation systems is dependent on accurate information on water availability (including seasonable and inter-annual availability, sustainable surface and groundwater volumes, and the interaction between surface and groundwater), existing and potential water use, water quality issues and impacts arising from existing or proposed water use, and who is using what amount of water and for what purpose. Within the SADC context there are considerable challenges in terms of both institutional capacity and the reliability of hydrological and water use information that beg the question of the efficacy of the allocation systems that are in place to both ensure adequate protection of vital human needs and the environment, as well as economic growth. This is in spite of the fact that the Revised Protocol requires State Parties to determine an appropriate balance between the socio-economic needs of their populations and the protection of shared watercourses.

The capacity of SADC states to allocate water between the sectors in pursuance of a coordinated growth policies, against the full water footprint of these sectors set against the water, food and energy nexus, appears even more questionable.

Current water allocation systems primarily deal with blue water i.e. ground or surface water that is abstracted for a particular use, and to some extent with grey water, or the pollution related implications of water use. Agriculture, on the other hand, uses large quantities of green water through rain fed agriculture, and in SADC as a whole, green water makes up some 92 % of the water footprint of agricultural goods traded within the region (CRIDF database). However, with the exception of South African legislation which has provisions for declaring Stream Flow Reduction Activities, there are no formal mechanisms to directly 'allocate' green water. Nonetheless, land use planning and the development of other infrastructure (transport and marketing) can influence placement of rain fed agriculture, thus acting as an indirect or proxy allocation mechanism.

In some countries in the region waste discharge permitting acts as an allocation function for grey water, but this does not deal with issues of pollution from dispersed sources such as agriculture.

World-wide, very few countries actively consider water footprints in their water allocation planning. The exceptions are Egypt and Israel who are deliberately following policies to import food, and hence Virtual Water, and rather allocate internal blue water to commercial and industrial users. China is also actively pursuing policies to reduce allocations to agriculture in favour of industrial and commercial users, while Saudi Arabia has deliberately reduced irrigation from groundwater within its borders, and has a policy of buying land in Africa to produce the food needed for its population.

5   Water Allocation Between States

The allocation of water between riparian states in shared river basins in SADC is guided by the provisions of the Revised SADC Protocol on Shared Watercourses (the Protocol),[1] which require that state parties use the waters of shared watercourses in a reasonable and equitable manner. This will become an increasingly complex allocation challenge as water stress grows, placing the overarching objective of the Protocol to, *"foster closer cooperation for judicious, sustainable and coordinated management, protection and utilization of shared watercourses, and advance SADC's agenda of regional integration and poverty alleviation."* (SADC Revised Protocol, Art. 2) even more in the spotlight.

The mechanisms for achieving reasonable and equitable use include:

- Establishing specific shared watercourse agreements and institutions (SWI);
- Advancing the equitable, reasonable and sustainable utilization of shared watercourses;
- Promoting coordinated and integrated environmentally sound development and management of shared watercourses;
- Promoting the harmonization and monitoring of policies and legislation related to shared watercourses;
- Notification of planned measures; and
- Promoting information exchange, capacity building, and research and development to support effective management.

Each of these mechanisms is elaborated in the Protocol. However, the following Articles are of particular relevance to the issue of allocation of water between states:

---

[1]With the exception of Zimbabwe which has not ratified the Protocol. However, Zimbabwe is Party to the ZAMCOM Agreement which includes similar provisions—and is in effect also bound by these principles.

- Article 3.8 (a)—borrowed from the 1997 UN Watercourses Convention—sets out the factors to be considered when establishing the reasonable and equitable use of water.
- Article 4.1 sets out the requirements for notification, information exchange, and negotiation/coordination among riparian States over planned measures that may have significantly adverse impacts on a shared watercourse.
- Article 4.2 requires member states to jointly and individually protect and preserve transboundary watercourse ecosystems and the aquatic environment, and to prevent, reduce and control pollution and environmental degradation of the watercourse (including the introduction of invasive species).

State parties are required to take steps to harmonize their policy and legislation in this regard, and to consult with each other on the setting, monitoring and enforcing of joint water quality standards and practices to address point and non-point source pollution.

- Article 4.3 requires member states to enter into consultations on the joint management of a shared watercourse, including the establishment of shared management institutions. States must also co-operate the regulation of flows and participate on a reasonable and equitable basis in the construction and maintenance or defrayal of costs of regulation works and to protect and maintain installations.

Member states are required to jointly take all appropriate measures to prevent or mitigate conditions that may be harmful to another riparian state, whether resulting from natural causes or human conduct, including regulating the actions of persons within their respective territories to prevent pollution or harm through the establishment and implementation of permitting/licensing systems.

While the Protocol defines a 'watercourse' to include groundwater implementation has, to date only focused on surface water.

In advancing the equitable and reasonable utilization of shared watercourses, the Protocol requires that state parties take a number of factors into account, summarised as follows:

- Physical or natural elements of the basin:

  - The **length of the river** lying in or adjacent to the regions;
  - The **area of the basin** lying within the territory of the regions; and
  - The **contributions made to the runoff** by the regions.

- Social or human needs and economic dependency on water:

  - **Water demands** exerted by the economy;
  - **Population dependent** on the shared waters;
  - **Extent** of that dependency; and
  - **Vital human needs**, the water required for basic human needs like drinking and sanitation.

Generally the courts have, in disputes between administrative regions over water, tended to favour the social and economic dependency on blue water, although sometimes adjustments are made to accommodate the physical and natural elements (McIntyre 2013). However, while this has to date not been considered in SADC, green water availability and its contribution to the economy, may affect the extent of the dependency on blue water resources. The consideration of the full water footprint could therefore result in different approaches to water allocations between states in a transboundary basin. For example, countries with higher rainfall could be seen as being less dependent on irrigation (blue water) and hence could have water allocations limited in favour of countries with less rainfall. Similar arguments have been made in the Cauvery Basin in India, where the downstream state argued that the upstream state had the advantage of a double monsoon season.

6 Basin Level and Bilateral Agreements

At the basin level, either basin-wide and/or bilateral agreements may determine the actual conditions for water allocation between signatories of these agreements. Any revision of these agreements requires a negotiated process between the relevant countries, which is a complex and time-consuming process. These agreements are seldom responsive to changing conditions, such as climate change, particularly where firm cross border flows have been agreed, although many include 'escape clauses' allowing deviation in cases of 'severe drought'. In many basins firm cross border flow agreements have proven very difficult to implement and sustain— particularly in times of water stress (Speed et al. 2013). However, there may be scope for shifting the way countries manage transboundary waters, towards a shared drought and flood risk approach using the infrastructure and blue, green and grey water allocations across the whole shared basin.

Table 1 summarises the basin level agreements in place in the transboundary basins of the region. What this does not capture is the many and varied bilateral agreements that are also in place in these basins. In many cases the actual details of water sharing are captured in the other (mostly) bilateral agreements rather than in the basin wide agreements which tend to be enabling rather than prescriptive.

However, some basin-wide agreements are more comprehensive, delegating more authority to the transboundary river basin organisations and that including more procedural and substantive obligations on the riparian states in line with the Protocol. For those agreements that pre-date the Protocol, various actions have been taken to broaden the scope of authority of the river basin organisations to facilitate more meaningful cooperation on substantive issues without altering the agreements themselves.

The Zambezi Watercourse Commission (ZAMCOM) agreement is the most recently ratified agreement in the region and includes requirements for national level policy and legislative harmonization across the basin. Many other

Table 1 Summary of agreements in place in transboundary basins in the SADC region

Basin	Agreement	Date agreement signed	Date agreement entered into force
Kunene	Agreement between the Government of the People's Republic of Angola and the Government of the Republic of Namibia on General Cooperation and the Creation of the Angolan-Namibian Joint Commission of Cooperation	18 September 1990	18 September 1990 (self-executing)
Okavango	Agreement between the Governments of Angola, the Republic of Botswana and the Republic of Namibia on the Establishment of a Permanent Okavango River Basin Water Commission (OKACOM)	15 September 1994	15 September 1994 (self-executing)
Orange-Senqu	Agreement between the Governments of The Republic of Botswana, the Kingdom of Lesotho, the Republic of Namibia and the Republic of South Africa on the Establishment of the Orange-Senqu River Basin Commission (ORASECOM)	3 November 2000	
Inco-Maputo	Tripartite Interim Agreement between the Republic of Mozambique, the Republic of South Africa and the Kingdom of Swaziland for Cooperation on the Protection and Sustainable Utilization of the Water Resources of the Incomati and Maputo Watercourses	29 August 2002	
Umbeluzi	Agreement between the Republic of Mozambique and the Kingdom of Swaziland on the establishment of a Joint Water Commission	1986	

(continued)

Table 1 (continued)

Basin	Agreement	Date agreement signed	Date agreement entered into force
Limpopo	Agreement between the Republic of Botswana, the Republic of Mozambique, the Republic of South Africa and the Republic of Zimbabwe on the Establishment of the Limpopo Watercourse Commission	27 November 2003	November 2011
Pungwe, Busi and Save (and potentially others shared between Mozambique and Zimbabwe	Agreement establishing the Joint Water Commission for the Pungwe	2002	December 2002 (self-executing)
Zambezi	Agreement on the Establishment of the Zambezi Watercourse Commission	13 July 2004	June 2011
Rovuma	Agreement Establishing the Rovuma/Ruvuma Joint Water Commission	2006	
Congo	Accord Instituant un Régime Fluvial Uniforme et Créant La CICOS	6 November 1999	23 November 2003
Lake Tanganyika	Convention on the Sustainable Management of Lake Tanganyika	2003	September 2005

agreements,[2] however, only establish the basin-wide commission as an advisory body. Harmonized national standards, instruments and mechanisms for basin-wide planning, information and data collection and management, water quality, and impact assessment procedures have only been established in a few basins, all of which would assist in improved understanding of allocation of water in transboundary basins.

Basin studies have been done in the majority of the basins in SADC, and a large percentage of these have formed the basis for the development of basin management strategies. However, there are a limited number of comprehensive basin management plans in existence for shared watercourses. While there are some in progress, a number of factors (most notably the requirement for approval from each of the members states of the basin) mean this process can take some time.

---

[2]LIMCOM, ORASECOM, OKACOM, and the Joint Water Commission (Mozambique and Zimbabwe).

Baseline studies have been completed for at least 11 of the transboundary basins in the region, while three management plans have been concluded (Congo, Inco-Maputo and Zambezi), two management plans are in progress (Orange-Senqu and Limpopo), and there are an additional five basin strategies or strategic action plans that may be considered as strategic plans (rather than strictly as management plans). However, the extent to which these plans have rigorously interrogated the requirements for reasonable and equitable use is debateable. As such, the extent to which these plans could be considered as inter-state water allocation mechanisms is uncertain, and would depend to a large extent on the way they were adopted, and the powers and functions that have been afforded to the commission. Certainly, it may be argued in some cases that 'approval' of basin wide plans by all riparian states which note plans for increased water use reflects an agreed water allocation. However, the enforceability of the plans in this regard is highly unlikely.

7 Water Allocation and Re-allocation

The issue of water allocation between sectors is particularly pertinent where a basin is stressed, is approaching closure, or is already closed. In many of these cases, after water efficiency measures have been accommodated, continued growth will require the reallocation of water between sectors, or the development of new water storage or transfer infrastructure. Recognising that 'new' water delivered through additional storage or transfers may become increasing expensive, and (in the face of increasing regional water stress) difficult to beget on a political basis, water managers may consider the importation of Virtual Water in agricultural products, rather than to produce them locally. This would have to be effected through shifting allocations between water use sectors, primarily to reduce the allocation to irrigation in favour of uses that provide greater income and jobs per drop—opting for more optimal financial, economic and environmental solutions. This approach remains nonetheless a legally and socially contentious approach.

The reality of the SADC context is that there is often an unregulated and informal transfer of water between sectors. For example, many cities and towns use more water than they are formally entitled to, not as the result of a formal allocation process but through increased physical withdrawals from surface and groundwater systems. The same pertains to many irrigation farmers. In most cases the 'user' that suffers is environmental flows, or weaker downstream users. Thus, in the face of weak regulation, allocation is often determined by the physical ability to abstract water rather than by the formal allocation system. This begs the question of how effective allocation systems are in the SADC region and the potential impact this may have on regional cooperation—particularly in the face of severe drought and/or climate change. For example, climate change may reduce outflows from Lake Malawi, but not to the extent that users within Malawi would be compromised. But the impacts on Mozambique and the Zambezi delta may be considerable. The extent to which Malawi's water allocation processes could accommodate these impacts,

Table 2 Progress on the implementation of management instruments for IWRM

Country	IWRM Planning	Allocation mechanisms	Monitoring and information
Angola	some, but limited achievements	little or nothing achieved	some, but limited achievements
Botswana	some, but limited achievements	some, but limited achievements	some, but limited achievements
Lesotho	substantial achievements or progress	some, but limited achievements	some, but limited achievements
Madagascar	some, but limited achievements	some, but limited achievements	some, but limited achievements
Malawi	some, but limited achievements	some, but limited achievements	some, but limited achievements
Mauritius	some, but limited achievements	some, but limited achievements	some, but limited achievements
Mozambique	some, but limited achievements	some, but limited achievements	some, but limited achievements
Namibia	some, but limited achievements	some, but limited achievements	some, but limited achievements
Seychelles	some, but limited achievements	some, but limited achievements	some, but limited achievements
South Africa	some, but limited achievements	some, but limited achievements	some, but limited achievements
Swaziland	some, but limited achievements	some, but limited achievements	some, but limited achievements
Tanzania	some, but limited achievements	some, but limited achievements	little or nothing achieved
Zambia	some, but limited achievements	little or nothing achieved	some, but limited achievements
Zimbabwe	some, but limited achievements	some, but limited achievements	some, but limited achievements

- = little or nothing achieved
- = some, but limited achievements
- = substantial achievements or progress

Towards a Water Secure Africa—Progress in IWRM in Eastern and Southern Africa (GWP-SA 2009)

reducing allocations to sugar irrigation, in the face of potential concerns raised by Mozambique is questionable. This particularly in light of the fact that the export of sugar derived from irrigated cane is a major foreign exchange earner for Malawi (see Chapter "Quantifying Virtual Water Flows in the 12 Continental Countries of SADC"), but is also directly pertinent to their commitments under the Protocol.

In 2009, GWP-SA conducted a survey on progress in the implementation of IWRM in the Southern and Eastern Africa. As Table 2 indicates, there were, at that point, limited achievements in relation to allocation mechanisms in the region, and similar weaknesses in terms of monitoring and information, which is a key element of implementing effective allocation. While there has been progress since then, a recent review of implementation of the SADC RSAP III revealed ongoing weaknesses in institutional capacity and monitoring and information. The issue of allocation mechanisms was not considered in this review.

Thus, the context of the allocation of water between sectors and users in the SADC region is one of weak institutional capacity and limited information.

Moreover, Muller postulates hydro-centric versus hydro supporting approaches, suggesting that there is little evidence that hydro-centric approaches have successfully driven development. It is therefore perhaps unrealistic to propose that a process of re-allocating water between sectors per se would be sufficient to shift planners away from large scale storage or transfer infrastructure as a solution to water stress. However, Virtual Water thinking may place powerful economic and politically savvy alternatives on the table. Done timeously, this may shift longer term planning in time to exert a gradual influence and shift.

8 Prioritisation of Water Use and Assurance of Supply to Address Variability

Most countries in the region have some degree of prioritisation of water use, particularly for water for basic human needs and for the environment. For example, Zambia prioritises water for domestic, livestock and urban use, but then affords all other uses the same priority. In the case of South Africa, this is taken one step further in the National Water Resource Strategy 2 (NWRS-2) which sets out a prioritisation of all water use and which should guide sectoral water use allocations at the national and basin level. This prioritisation also underpins the assurance of supply principle, as is highlighted in Table 3.

This allocation system specifically accommodates variability in supply, while seeking to maintain the economic viability of the water use, or its role in the country's economy. Varying the assurance of supply in this way frees up water for further allocation. For example, in the Crocodile sub-catchment in South Africa, a maximum of 342 million m^3/year is available for allocation 100 % of the time, this increases to 427 million m^3/year at a 95 % assurance, 463 million m^3/a, at a 90 %, and 499 million m^3/a at an 85 % assurance.

A Virtual Water perspective can also play a role in establishing fair assurance of supply rules. Users that can rely on higher rainfall and hence green water contributions may be able to accommodate reduced assurance while still remaining economically viable. An improved understanding of the water footprint of the various water users, set against their contributions to employment and the economy may place water managers in a better position to build climate resilience through water allocations.

Table 3 Assurances and curtailments considered viable under South African circumstances

User sector	Assurance (%)	Maximum curtailment (%)
Primary domestic needs	100	0
Strategic	99	5
Industrial	98	20
Urban	95	30
Irrigation		
– *High value*	95	30
– *Medium value*	90	50
– *Low value*	80	70
– *Opportunistic*	70	100

9   Challenges for Considering Virtual Water in Water Allocations

Apart from the proposals outlined in the previous sections, there are a number of specific issues that need to be examined when considering the potential role for, and practicality of, Virtual Water in water allocation systems:

- To what extent are blue and green water adequately considered in the cost driver and allocation picture?
- Do decision-makers have the information, framework and tools for considering the possibilities of using green water in one area rather than blue water in another area?
- To what extent can the choice between infrastructure, further developing groundwater sources, or Virtual Water be informed by a water footprinting approach?
- How do Virtual Water principles influence issues of direct cost, job creation, risk and political imperatives?

It is worth noting, at this point, that apart from South Africa, Zimbabwe, and to some extent Namibia, the region is poorly served with major storage infrastructure, making it extremely vulnerable to even short term drought. Drought, and the water, food and energy nexus also plays into the scenario of hydropower development in the region. In the Zambezi basin, states will have to increasingly make trade-offs between irrigation and hydropower. Similarly, states reliant on hydropower (either imported or locally produced) from basins potentially prone to variable water availability, may also need to consider their strategic risks or to consider alternative sources of electricity in droughts. The consideration of Virtual Water in allocation decisions therefore has to be underpinned by a broader understanding of the full water footprint in the national and regional economy. The following sections address these challenges from the SADC perspective.

10   Economic Accounting for Water

Currently water allocation is based on the use of hydrological modelling to determine water availability, and to allocate this water to various economic sectors according to demand and levels of priority of those sectors in the economic development framework of the country. In many cases, basic water supply has the highest priority. The allocation sometimes involves drought rules that determine which sectors receive priority and which are most heavily curtailed in times of drought or water shortages. This approach ensures the allocation of water to those sectors that are envisaged of being high priority. However, according to the SADC project on Economic Accounting for Water, "*it does not lead to* **pareto optimality**— *that is allocation of the scarce water resources in such a way that maximises the social welfare benefits for the River Basin. In order to achieve pareto optimality information is required on* **Value Added**, **Water Productivity**, *and* **Water Use Intensity** *and this information is usually not included in current models which are used by river basins to allocate water*" (*SADC* Water Accounting).

However, the economic accounting for water still does not accommodate the full impact of water on the economy, and only addresses blue water, thus missing some 80–90 % of the water in agricultural products that is derived from green water.

11 The Nexus Approach

It is very clear, in the SADC context, that there are regional and national benefits to be derived from appropriate siting of energy generation and crop production (rain fed and/or irrigated). Where possible the siting of irrigation abstractions below hydropower plants can allow for the dual use of water and hence deriving greater benefits from the available resources. Similarly, the careful conjunctive operation of hydropower plants and irrigation abstractions can hold net benefits across the basin (World Bank 2010). Given adequate transmission capacity, electricity trading from basins with significant runoff at a high assurance, like the Congo River, can meet significant regional demands, but SADC states have shown a reluctance to forgo sovereign electricity security—allocating water perhaps preferentially to irrigation (see Chapter "Electrical Power Planning in SADC and the Role of the Southern African Power Pool").

Moreover, on a national basis, hydropower dependent countries may have to make difficult allocation decisions between hydropower and irrigation. While countries like Zambia earn considerably more in agriculture exports than electricity exports, a lack of power has much greater impacts on the economy, while irrigation itself often relies on electricity. However, the sugar industry may provide for its own electricity through bagasse, and this may be an important consideration in countries where the sugar industry is an economic mainstay, like Zambia and Malawi. Appropriate water allocations and assurance of supply in these countries will therefore also be influenced by electricity demands and possible surpluses, variable runoff (and climate change) and the viability of purchasing regional electricity surplus through the Southern African Power Pool (SAPP). Conversely there are transnational risks arising from the off-shoring food production, and the potential costs (and foreign exchange implications) of importing food. Regional integration has not yet reached the level at which SADC states see national interest within the context of the general regional interest.

12 Benefit Sharing as an Allocation Approach

The concept of benefit sharing in transboundary basins has been widely talked about. Benefit sharing is "the process where riparian countries cooperate in optimising and equitably dividing the goods, products and services connected directly or indirectly to the watercourse, or arising from the use of its waters" (Phillips and Woodhouse 2009).

The concept of benefit-sharing is closely aligned to the concept of Virtual Water, although it is generally applied only within the confines of one basin, while Virtual Water can be regional or even global in scope. A benefit-sharing approach enables one to move from a focus on the physical volumes of water to be shared, to an analysis of the economic, social, political, and environmental benefits to be derived from the water use. In theory, at least, it allows allocation of water to generate the optimal social, economic and environmental benefits in the basin. The challenge, however, is enabling riparian states to understand the potential for a positive-sum outcome from this approach, rather than the constrained approach arising from a volumetric allocation, through the Virtual Water and economic accounting lens. For example, the Lesotho Highlands scheme is an often quoted example of such a benefit sharing approach. However, this was not driven on the basis of the equitable sharing of the full benefit of the use of water in South Africa, versus Lesotho, but rather on the incremental difference in the cost of infrastructure built in Lesotho, as opposed to on the Orange River once it entered South Africa.

Taking a truly benefit sharing approach (as with a Virtual Water approach) requires a quantification of the costs and benefits associated with using the water from a particular resource, and the application of agreed valuation techniques to assess the full economic and social value of the costs and benefits. There has been some work done around this by the SADC Water Division through a programme on Economic Accounting of Water, which has been piloted in the Orange-Senqu river basin.

13 Challenges of Accurate Information

One of the major challenges of any water allocation system, whether between states, between sectors, or to individual water users, is the accuracy of the hydrological and water use data on which it is based. Unfortunately, in the SADC context, such data is limited, and not particularly reliable. The hydrological and meteorological monitoring systems in the region are under-developed and poorly maintained, resulting in limited data generation. In addition, records of actual water use are also weak or non-existent in most places.

Weaknesses in hydrological monitoring are compounded by climate change and the difficulty of linear projections based on historical patterns which are no longer valid. From the climate change perspective, there is weak downscaling of climate change models in the SADC region apart from South Africa, leaving high levels of uncertainty in relation to the hydrological and meteorological impacts of climate change.

Applying a Virtual Water and water footprint layer to this may result in further uncertainty and inaccuracies.

14 Enabling Environment to Use Water

A further challenge that exists in the SADC region is weaknesses in the enabling environment in order to be able to use water for economic development, including export of agricultural produce through adequate transport infrastructure and energy provision to the more remote areas.

Taking this issue further and looking at the trade of Virtual Water in relation to agriculture, as captured in Table 3, an interesting picture begins to emerge. Water allocations are focused on apportionment of blue water, with 80–90 % of blue water used in irrigation in developing countries. It is, therefore, worth examining the issue of blue water content in agricultural imports and exports in the region. States with high blue water content in their agricultural exports, and where those exports are significant earners relative to the GDP, and where there is little storage, are vulnerable to short term changes in runoff, and potentially the multiplier effects of increased temperatures and reduced rainfall on runoff. Conversely, those economies that are heavily green water dependent may be vulnerable to even a month with less rain than expected. Indeed, in many African countries reduced economic growth tracks rainfall patterns.

On the other hand, countries that are dependent on agricultural imports are at risk from droughts in the countries from which they are importing, or even global food price fluctuations. For example, severe drought in the Former Soviet Union countries in 2012 forced up global wheat prices and with that overall food prices. The allocation of water during times of drought in particular, and the implications of water allocation and curtailment decisions on the economy, therefore are particularly relevant.

Malawi and Swaziland, and to some extent Zambia appear to be vulnerable in this regard (Table 4). Their trade surpluses in agricultural goods an important contributor to GDP yet, apart from Swaziland, they have virtually no storage (and much of Swaziland's 'storage' lies in South Africa). Climate vulnerability may therefore further constrain economic growth in these countries. Breaking out of this trap will require a broader enabling environment through both improved storage and transport infrastructure (spreading the agricultural base to wetter regions) and accommodating shorter term droughts, improved water governance and the introduction and policing of variable assurance of supply across the irrigation and energy sectors, and a diversified economy. A lack of this enabling environment is likely to be the biggest challenge to effectively building Virtual Water thinking into water allocations. Certainly a catch 22 may arise where Virtual Water thinking is

Table 4 Virtual water imports and exports in SADC per country

Country	Renewable water (m³/cap)	Storage (m³/cap)	GDP U$/ PPP (U$ billion)	GDP (U $/cap)	Virtual water imports in agricultural goods (%)			Virtual water exports in agricultural goods (%)			Agric export earning from blue water (U$ illions)	Agricultural trade surplus/deficit (U$ billion)
					Blue	Green	Cost $ billion	Blue	Green	Earn $ billion		
Angola	7333.82	468.40	114.15	6006.34	9	82	2.4	0	100	0.015	0	−2.385
Botswana	1208.03	221.00	14.50	16,104.91	11	81	0.5	4	96	0.92	37	0.42
DRC	27,220.00	0.76	17.20	415.34	14	77	0.45	0	100	0.038	0	−0.412
Lesotho	2576.97	1272.00	2.45	1931.21	7	83	0.02	12	86	0.005	1	−0.015
Malawi	1044.15	2.63	4.26	753.38	14	76	0.14	10	84	0.9	90	0.76
Mozambique	4080.33	3165.00	14.24	1007.23	14	78	0.43	13	86	0.52	68	0.09
Namibia	2777.76	299.70	13.07	7442.34	10	76	0.6	14	85	0.56	78	−0.04
South Africa	868.56	601.70	384.31	11,021.02	9	84	4.73	13	82	4.97	646	0.24
Swaziland	2177.93	479.50	3.74	5161.14	10	80	0.023	40	60	0.24	96	0.217
Tanzania	1812.12	2187.00	28.24	1574.78	12	80	0.88	2	97	1.11	22	0.23
Zambia	5882.44	250.00	20.59	1682.86	5	90	0.35	8	88	0.88	70	0.53
Zimbabwe	917.75	400.00	9.80		9	82	1.13	15	71	1	150	−0.13
SADC average	4824.99	778.97	52.20	4827.32	6	91	11.7	10	86	15.7		

required to make best use of variable water availability in growing the economy of these countries, but this cannot be introduced effectively because the economy is not large enough provide an appropriate enabling environment.

15 Allocation Mechanisms that Function Within State Capabilities

The previous section underpins one of the key constraints faced with respect to successful water allocation mechanisms; the capacity of the state to effectively implement them. Not all of the allocation systems contained in law are implementable within the human, financial and data resource limitations in the region. All allocation systems, whether administrative or market based, require a minimum platform of infrastructure and systems, particularly information on existing and current water use, and water availability.

In addition, allocation systems need to be able to deal with inter-annual variability and the impacts of climate change. This requires adaptive and flexible systems, and the ability to analyse changes in rainfall and run-off. While there are considerable hydrological skills and capacity in most SADC countries, it is likely that sectoral and even individual allocation is taking place organically, rather than systematically due to a lack of institutional resources and the high transactional costs of hydro-centric approaches. For example, municipal water use in many countries is increasing not through a formal allocation of water, but simply through increased abstraction, and expansion of the urban boundary.

16 Land and Water Grabs

One of the ways that allocation of water takes place is through the purchasing of land with associated water entitlements. There is increasing concern, that large scale 'land-grabbing' by more wealthy states, particularly from outside the region, may result in local people being disposed of their right to land and water.

After the recent food price crisis, several countries and companies bought land in Africa, and foreign land purchases in central Africa now totals some 16 million hectares (although this number is in some dispute). In some cases, this has contributed to the local economy by creating jobs and providing ready markets for agricultural products, albeit outside of the region (hence representing a net Virtual Water loss to the region). However, in many cases little has been done to develop and use the land, thus trapping the potential blue and green water associated with that land. This is a form of virtual land and water allocation but without sufficient regulation or necessarily assessment of the regional impacts.

Nonetheless, as indicated in Chapter "Quantifying Virtual Water Flows in the 12 Continental Countries of SADC", SADC is a net Virtual Water importer, yet earns more foreign exchange in its Virtual Water exports.

17 Conclusion

Muller's (Chapter "Virtual Water and the Nexus in National Development Planning") observation that hydro-supportive approaches, allocating water in support of demands, rather than hydro-centric, using water allocations to drive demands, is most likely to be the dominant paradigm is broadly supported by the author.

However, in this regard, Wichelns notes that: "*The virtual water metaphor, while not a sufficient criterion for determining optimal strategies, still serves an important role in gaining the attention of public officials. Once that is accomplished, the discourse can be extended to include consideration of opportunity costs and comparative advantages, as strategies are determined and policies are selected*" (Wichelns 2004).

Through the Virtual Water lens, policy makers can examine the issue of scarcity of water and optimal allocation options in and between countries and sectors. However, the current inter-country agreements are largely volumetric in nature, and a move towards a consideration of Virtual Water approaches would require a considerable paradigm shift. Similarly, SADC states and transboundary basin Commissions do not seem ready to thoroughly examine reasonable and equitable use principles even on a blue water, let alone water footprint basis.

At the level of inter-sectoral allocation, it is not clear that the allocation systems currently in place, or the monitoring and information systems, are sufficiently well-developed to accommodate effective allocation of water through a Virtual Water lens, in spite of the potential that the concept offers.

Nonetheless, similarly to Wichelns, it is the opinion of the author that looking at water allocations the Virtual Water, water footprint, and nexus lens places other options on the table. These options may indeed influence management decisions away from solely examining large infrastructure solutions to variability and stress in water availability. The development of powerful economic and politically savvy messages, as outlined in Chapter "Embedding the Virtual Water Concept in SADC", would lend further weight to influencing regional national planning processes.

References

Care Climate Change (2015) Available at: http://www.careclimatechange.org/files/reports/Implications_drought_risk_world_7.jpg. Accessed 03 June 2014
Global Water Partnership South Africa (2009) Improving Africa's water security. Progress in integrated Water Resources Management in Eastern and Southern Africa. Available at: http://

www.gwp.org/Global/ToolBox/References/Improving%20Africa%20Water%20Security%20 (GWP,%202009).pdf. Accessed 11 Jan 2016

McIntyre O (2013) Factors relating to equitable utilisation of shared freshwater resources. Water Int 38(2):112–129

Mott-Macdonald EM (2008) Integrated Water Resources Management Strategy and Implementation Plan for the Zambezi River Basin. Available at: http://www.zambezi commission.org/index.php?option=com_content&view=category&layout=blog&id=16&Itemid= 178

Phillips D, Woodhouse M (2009) Transboundary benefit sharing framework: training manual (version 1). Prepared for benefit sharing training workshop. Addis Ababa

Southern African Development Community (2011) SADC statistics yearbook. Available at: http:// www.sadc.int/information-services/sadc-statistics/sadc-statiyearbook/. Accessed 15 June 2014

Speed R, Li Y, Le Quesne T, Pegram G (2013) Basin water allocation planning. Principles, procedures and approaches for basin allocation planning. Available at: http://www.adb.org/ publications/basin-water-allocation-planning. Accessed 07 June 2014

Wichelns D (2004) The policy relevance of virtual water can be enhanced by considering comparative advantages. Agric Water Manag 66:49–63

World Bank (2010) The Zambezi River Basin: a multi-sector investment opportunities analysis. Available at: http://documents.worldbank.org/curated/en/docsearch/report/58404. Accessed 07 June 2014

Electrical Power Planning in SADC and the Role of the Southern African Power Pool

Simon Krohn, Simbarashe Mangwengwende and Lawrence Musaba

Abstract Recognising the benefits of regional integration, SADC countries established the Southern African Power Pool (SAPP) through an intergovernmental Memorandum of Understanding signed in August 1995, SAPP's mandate is to provide non-binding regional masterplans to guide electricity generation and transmission infrastructure delivery, with countries retaining the right to develop and prosecute their own national plans. Although SAPP masterplans have been able to demonstrate considerable financial savings and other benefits of regional cooperation, countries have instead continued to develop plans for achieving electricity self-sufficiency. SAPP interactions have demonstrated that regional cooperation requires the adoption of multi-sector planning approaches to establish credible demand forecasts, and a multi-criteria approach to selection of project options going beyond identification of regional least cost options, to incorporate sovereign electricity security and other national interests. In particular, regional plans must be able to demonstrate, from each country's perspective, the equitable sharing of benefits of integration, while addressing the legitimate security and social concerns.

Lawrence Musaba—Deceased

S. Krohn (✉)
Simon Krohn Consulting Pty Ltd, Hobart, Australia
e-mail: snkrohn@gmail.com

S. Mangwengwende
Zambezi Hydro Power Company, Harare, Zimbabwe
e-mail: mangwe.simba@gmail.com

L. Musaba
Director Southern African Power Pool Coordination Centre, Harare, Zimbabwe

© Springer International Publishing Switzerland 2016
A. Entholzner and C. Reeve (eds.), *Building Climate Resilience through Virtual Water and Nexus Thinking in the Southern African Development Community*, Springer Water, DOI 10.1007/978-3-319-28464-4_6

133

1 Introduction

This chapter addresses energy sector planning issues focussing on the electricity supply industry which has the most significant impact on the SADC region's carbon footprints, and is an important component of the regional water footprint. Electricity planning takes place at both national and regional levels through the Southern African Power Pool (SAPP). The SAPP was established in August 1995 at the SADC Summit held in South Africa when an inter-governmental memorandum of understanding (IGMOU) was signed. A revised IGMOU was signed on 23 February 2006 to allow for new SADC members and new enterprises created as a result of the restructuring of the power sectors of the member countries. Current membership of SAPP comprises the national electricity utilities of the 12 continental members of SADC, an independent power producer (IPP) and independent transmission company (ITC) from Zambia. An IPP and ITC from Mozambique are observer members.

2 Mandate of the SAPP

The SAPP mandate is guided by the SADC Protocol on Energy signed in Maseru on 24 August 1996 which is the principal policy document governing the SADC energy sector. The protocol defines the guidelines for co-operation in the electricity sector, which include the development and updating of a regional electricity masterplan, development and utilisation of electricity in an environmentally sound manner, and emphasising the need for universal access to affordable and quality services.

The SAPP IGMOU expresses the clear intention of member countries to enhance regional cooperation in power development and trade and defines the basis for the establishment of the power pool as the need for all participants to:

(a) Co-ordinate and co-operate in the planning, development and operation of their systems to minimise costs while maintaining reliability, autonomy and self-sufficiency to the degree they desire;

(b) Fully recover their costs and share equitably in the resulting benefits, including reductions in required generating capacity, reductions in fuel costs and improved use of hydro-electric energy; and

(c) Co-ordinate and co-operate in the planning, development and operation of a regional electricity market based on the requirements of SADC member states.

Consistent with the principle that member countries are free to choose the degree of autonomy and self-sufficiency desired, a separate Inter-Utility Memorandum of Understanding (IUMOU) states that the regional generation and transmission masterplan shall be "purely indicative and shall not create an obligation upon members to comply" (clause 12.5.2 (vi), SAPP IUMOU—April 2007). Any regional generation and transmission masterplan must therefore hold considerable regional and national benefit after satisfying the desired level of autonomy and self-sufficiency, in order for it to attract attention and investment.

3 Methodology

In order to introduce the Virtual Water and nexus concepts to policy and decision makers involved with the electricity sector the paper adopts the following approach:

- Provide information on current and projected power supply and demand situation and Virtual Water implications of electricity trades;
- Provide information on projected demand and the factors influencing that demand, as well as the regional and national options for fulfilling that demand and the factors driving the demand for electricity;
- Analyse the regional versus national planning challenges based on the experience of the SAPP in the discharge of its mandate;
- Outline a conceptual scenario for how Virtual Water and nexus concepts can help to fulfil the legitimate expectations of all stakeholders, including governments (and their development partners), regulators, utilities, private sector, civil society and consumers, especially the poor and marginalised.

4 Current Power Supply and Demand in SADC

The latest available electricity power and energy supply and demand statistics for the year ending 31 March 2013 published in the SAPP Annual Report are summarised in Table 1.

Table 1 SAPP power and energy demand (2013/14)

Country	Utility	Installed capacity (MW)	Dependable capacity (MW)	Peak demand* (MW)	Energy sent out (GWh)	Energy sales (GWh)
Angola	ENE	2028	1805	1333	5613	3427
Botswana	BPC	892	460	580	372	3118
DRC	SNEL	2442	1268	1342	7411	6688
Lesotho	LEC	72	72	138	486	488
Malawi	ESCOM	351	351	278	1809	1476
Mozambique	EDM/HCB	2308	2279	763	390	2380
Namibia	NamPower	501	392	635	1305	3648
South Africa	ESKOM	44,170	41,074	38,775	237,430	224,446
Swaziland	SEC	70.6	70	222	288.1	1018.6
Tanzania	TANESCO	1380	1143	898	3034	3770
Zambia	ZESCO/CEC/LHPC	2128	2029	2287	11,381	10,688
Zimbabwe	ZESA	2045	1600	2267	6951	7367
Total SAPP		58,388	52,543	56,821**	276,470	261,148
Total interconnected		54,628	49,244	54,312**		

Source SAPP Annual Report, 2013/14. *Includes estimate of suppressed demand; ** Includes reserves

The difference between the total interconnected capacity of 54,628 MW and the total capacity of 58,388 MW is due to the fact that three countries, Angola, Malawi and Tanzania are not yet interconnected. It is also important to note that the dependable capacity (49,244 MW) is lower than the peak and suppressed demand including required reserves for several countries and for the region as a whole. Suppressed demand is an estimate of additional consumption that would be on the public network if there were no supply constraints. It represents the demand of deferred investments or what is self-generated. The best practice reserve margin for a hydro-thermal system is about 15 % of the total peak demand of 49,563 MW which gives a total supply requirement of 56,821 MW. The region has therefore a dependable capacity shortfall of 4278 MW or close to 10 %. Reduced water inflows into Lake Kariba during the 2014/15 season worsened the capacity shortfall for Zambia and Zimbabwe by close to 700 MW. Many countries therefore have to resort to regular load shedding which adversely affects economic and social development.

For several countries the benefits of interconnection are evident in that their sales exceed their generation. However, the fact that they are now having to load shed because the supplying countries are unable to export enough to meet demand, has highlighted the need for importing countries to increase investment at national level in order to become more self-sufficient. The capacity of ongoing and committed projects in the short term for the different countries is highlighted in Table 2.

South Africa has nearly 80 % of the generating capacity, and 85 % of the energy generated in SADC is consumed in South Africa. The country also accounts for almost 60 % of the new capacity to be developed in the next four years. With coal accounting for 86 % of South Africa's generation capacity and hydropower accounting for the bulk of the remaining countries' generation these two technologies are the main sources of electricity in the region (Fig. 1). The Democratic

Table 2 MW capacity of committed short-term generation projects

Country	2014	2015	2016	2017	Total
Angola	204	0	1280	2271	3771
Botswana	150	0	0	0	150
DRC	0	580	0	240	820
Lesotho	0	0	35	0	35
Malawi	0	0	0	34	34
Mozambique	175	0	40	300	515
Namibia	0	0	15	0	15
South Africa	4836	1805	3717	1918	12,276
Swaziland	0	0	0	0	0
Tanzania	450	240	660	250	1600
Zambia	195	735	40	126	1096
Zimbabwe	0	15	0	1140	1155
Total	6026	3375	5787	6279	21,467

Source SAPP (Unpublished internal report, April 2014)

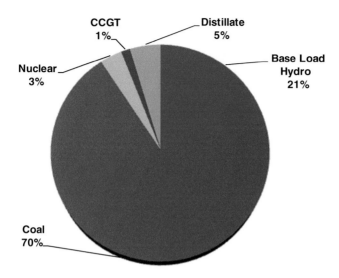

Fig. 1 Generation mix in SAPP. *Source* SAPP Annual Report, 2013/14

Republic of the Congo (DRC), Malawi, Mozambique and Zambia are almost entirely reliant on hydropower.

Projects completed in 2013 and due to be completed in 2014 comprise: 3269 MW of thermal generation, mainly coal in South Africa and Botswana and gas in Mozambique, South Africa and Tanzania; 1976 MW of hydro, mainly in South Africa and Zambia; 2112 MW of other renewables, mainly solar and wind in South Africa which also has an additional 30 MW nuclear. The generation mix for additional power added in these two years will be 44 % thermal, 27 % hydro and 29 % other renewables and nuclear.

5 Electricity Trading

Electricity trading in the SAPP is 99 % based on bilateral contracts of one month or longer duration and 1 % on short term day ahead market (DAM) (24 h duration). In 2012/13 there were 28 bilateral contracts but only 15 of these were active due to generation and transmission constraints (SAPP Annual Report 2013). The following Table 3 highlights the most active bilateral contracts for 2013. In Table 1 Zimbabwe is shown as one of the countries with a significant supply shortfall and yet it is noted that it is one of the large trading partners. This is a unique arrangement where NamPower (Namibian power utility) pre-paid for the power by providing financial support for major overhaul at Zimbabwe's Hwange Power Station, thereby increasing the output of a power station that would otherwise have been shut down. While Zimbabwe is still short of power and has to load shed, the

Table 3 Bilateral contracts in SAPP for 2013

Exporting country and utility	Importing country and utility	Capacity (MW)	Total (MW)
South Africa (ESKOM)	Mozambique (MOZAL)	950	1950
	Mozambique (EdM)	300	
	Swaziland (SEC)	250	
	Namibia (NamPower)	200	
	Botswana (BPC)	150	
	Lesotho (LEC)	100	
Mozambique (HCB)	South Africa (ESKOM)	1150	1670
	Mozambique (EdM)	270	
	Zimbabwe (ZESA)	250	
Zimbabwe	Namibia (NamPower)	150	150
Mozambique (EdM)	Swaziland (SEC)	50	135
	Botswana (BPC)	45	
	Namibia (NamPower)	40	
Total			3905

Source SAPP Coordination Centre (unpublished internal report, April 2014)

extent of the load shedding would have been worse without the NamPower support. It is a good example of the benefit of having the regional interconnected grid.

The DAM is a competitive market administered by the SAPP Coordination Centre which seeks to utilise the uncommitted generation and transmission capacity to help utilities cope with their day to day operational requirements. The sale and buy bids received are matched to establish a market clearing price but the actual trading depends on the available transmission capacity. Between December 2009 and March 2013 there was a total of 230,131 MWh of matched sale and buy bids but only 62,154 MWh (27 %) could be traded.

6   Barriers to Proposed Transmission Projects to Relieve Trading Constraints

The priority transmission projects to facilitate improved trading fall into two categories (a) the interconnection of Angola, Malawi and Tanzania to complete the interconnection of all member countries, and (b) the reinforcement and upgrading of transmission corridors used for wheeling power. The first category projects include internal transmission reinforcements in Zambia and a 400 MW Zambia-Tanzania interconnector which is envisaged to be extended to Kenya, 300 MW Mozambique (Cahora Bassa) to Malawi interconnector, 600 MW DRC to Angola interconnector and 400 MW Angola to Namibia interconnector. The second category projects included the 600 MW Zimbabwe/Zambia/Botswana/Namibia (ZIZABONA) interconnector, reinforcements of the Central Transmission Corridor

(CTC) in Zimbabwe, and transmission upgrades in South Western Zambia and North Western Botswana.

Experience has since proved that project implementation can be very slow notwithstanding obvious technical and financial benefits. Nearly twenty years have passed without completing the interconnection of the remaining three countries, namely Angola, Malawi and Tanzania. The interconnection of Angola to the power pool was originally envisaged as part a Western Corridor (Westcor) designed to supply power from the Inga scheme in the DRC all the way to South Africa through Angola, Namibia and Botswana. A special purpose company was created with an office in Botswana. The project was however aborted when the countries failed to agree on the formula for sharing benefits and other conflicting interests (see Text Box 1).

Text Box 1: WESTCOR—Project Collapses Over Lack of Shared Benefits Expectations

The Westcor project was conceived by the SAPP in 2002 with the objective of developing Inga-III in the DRC to supply power to South Africa with the three countries in between the DRC and South Africa, namely Angola, Namibia and Botswana also taking a share of the power. The project was not only serving to interconnect Angola to the SAPP grid but would also facilitate the establishment of a telecommunication system to link the five countries. With five participating countries this was an ideal regional cooperation project. In 2003, inter-governmental and inter-utility memorandums of understanding were signed by the Energy Ministers and national utilities of Angola, Botswana, DRC, Namibia and South Africa. A joint development agreement was signed and a shareholders' agreement for a company to serve as the Special Purpose Vehicle (SPV) for development of the project. The five utilities each paid US$100,000 into the establishment of the SPV. Botswana was selected to host the SPV and an office was established in Botswana.

Things went downhill as the DRC stopped attending meetings without formal notice. The following are *suspected* to be some of the reasons:

(i) The DRC felt that the signed MOUs were no longer binding. In their understanding once MOUs are signed the signatories should commit funding to the project which was not done in this case.

(ii) The DRC also felt that they needed more royalties from the Inga-III since the land and water all belonged to the DRC and they were against the idea of sharing benefits equally with other Members.

(iii) On top of being a member of the SAPP, like the other Members of Westcor, the DRC was also a member of other power pools like CAPP and EAPP. The other power pools which are even closer to Inga also wanted a share in Inga-III as the DRC had earlier promised them. This was therefore going to be a challenge for DRC to fulfil all the obligations.

In hindsight the signing of the IGMOU at Ministerial level and not Presidential level could have undermined the project. However, several pre-SAPP projects such as the Interconnectors from Zimbabwe to Cahora Bassa and to Matimba through Botswana were successfully implemented through agreements signed at Ministerial level.

The above reasons were enough for the Westcor project to collapse.

The Malawi-Mozambique interconnector progressed to the point of financial closure but still failed to take off for unrelated political reasons and perceptions of inequity in the sharing of benefits (see Text Box 2).

Text Box 2: Mozambique Interconnector—Politics Overrides Technical and Financial Feasibility

The Mozambique-Malawi Interconnection Project is a project that is aimed at Interconnecting Malawi, a non-operating member of the SAPP, to the SAPP grid via Mozambique.

In the year 2007/8, the project had reached financial closure and funding was available from both the World Bank and the Government of Norway. The World Bank had availed funding to both governments without the requirements for a power purchase agreement (PPA). The funding from Norway was to reduce the cost of the power to be supplied by Mozambique to Malawi and would act as a subsidy. Despite the fact that funding was available and the project would benefit both countries as well as the region, the project still collapsed due to some of the following reasons:

(i) The Government of Malawi had wanted to use the Shire River as a gateway to the Indian Ocean via Mozambique. The President of Malawi, President Bingu wa Mutarika at that time, had planned to establish an inland port in Malawi at the shore of Lake Malawi. Ships were to dock at this inland port from the Indian Ocean via Mozambique using the Shire River. The Government of Mozambique had refused this arrangement as no feasibility studies had been done to use the Shire River in this manner right across Mozambique. Because the Government of Mozambique refused to allow Malawi this request, the Government of Malawi also felt that they should refuse to accept the Mozambique-Malawi interconnector that would pass through Malawi and also supply the Northern part of Mozambique.

(ii) The Government of Malawi also concluded that Mozambique would be the largest beneficiary of the project as it was felt that the proposed tariff was too high and Malawi would be held captive by Mozambique and would not be in a position to access other funding to develop their own generation.

Other constraints in the implementation of power projects in the SADC region include the following:

- *Lack of political commitment to implement cost reflective tariffs*: The extent to which countries are committed to projects is demonstrated by their commitment to their sustainability and this in practical terms means having project beneficiaries pay prices that cover investment and on-going operation and maintenance costs. Current average tariffs in the region are within the range of 4.8–11.5 USc/kWh (SAPP Annual report 2013) and are not linked to long run marginal costs which are much higher than the sunk costs of existing plant. By comparison average tariffs for other sub-Saharan Countries is 13 USc/kWh and as high as 20 USc/kWh in West Africa.
- *Absence of a project driver or champion who is both resourceful and persuasive and who derives his mandate from the highest authority*: This is another factor that demonstrates weak political commitment to regional cooperation. Countries have or are considering self-sufficiency as the answer to their national energy security concerns.
- *Weak institutional frameworks and project preparation capabilities*: The lack of a harmonized framework across the region, which was of concern given the importance of these factors towards the success of credit enhancement techniques, and the generally poor relationships and lack of coordination between utilities and their regulators in some cases are significant barriers. Potential funders, (and some of the regulators and utilities) all have bemoaned the fact that projects are brought to them while they are still very rudimentary, and in need of a lot of project preparation/structuring/packaging work.
- *Dependence on PPAs to implement projects*: Most utilities in the region have very weak balance sheets; a number are technically insolvent and would not be deemed to be Going Concerns without continued government support. This makes it difficult for such utilities to attract finance to their projects. Eskom (South Africa) is by far the largest utility in the region. It is also one of the few that are credit rated. As a result of these two factors, potential funders have tended to insist that Eskom be not only the major off-taker, but also take up a lot of the risk on a project, by way of a PPA. On the other hand, the South African Government and Eskom seek to minimise risk by preferring to transact in South African Rand thereby passing on exchange risks to the project developers, hence the failure to conclude any PPAs.
- *The perception that dependence on other countries for electricity supply is a security risk*: Countries are pursuing electricity independence as far as possible, limiting their dependence on imports. This is not only due to the perceived security risk (or lack of sovereign control—being at the 'mercy' of countries which may renege on deals to meet their own needs), but also because of the perception that it would limit opportunities for national investments in energy generation and the associated job creation.

The cardinal principle in limited recourse funding is that the person who can best manage the risk takes the risk. For example, the private sector should take the technology risk, while government should guarantee the financial credit-worthiness of its utility and the risk of closing the gap between current tariffs and cost reflective tariffs required in PPAs. The region needs to realise that it is competing with other regions globally for private funds, and these will be invested where the governments are creating the right environment for such investment. It is a given that from the onset, the private sector, in general, do not want to take up much risk and they will try to pass as much of the risk to the utility or the government and seek a rate of return well beyond what they would get in developed countries. Accordingly, it will be important to ensure an appropriate risk sharing matrix is developed for the region, so that both the developers, governments and utilities are aware of what is acceptable, which will then facilitate standardisation of PPAs. The dependence on PPAs in the current funding models is certainly a factor, as is the dominance of Eskom as a supplier to the region. In this regard, further concerns stem from the negative impact of the downgrading of Eskom's credit rating on new projects.

7   Plans to Meet Future Demand

The plans to meet future demand are largely a function of the forecast and the criteria used to select options to meet the demand. The forecasts are done at national level and are then aggregated to give an indication of the regional demand. Since most of the national demand tends to be coincident in terms of the time of peak the regional demand is taken as the sum of the individual national demands. In practice there is some diversity that provides opportunities for meeting demand with reduced generating capacity (Table 4).

The following shortcomings in the demand forecasts are evident:

- The power and energy demand forecasts are a simple extrapolation of the current situation in each country. This does not take into account the universal access to modern energy by 2030 which is the United Nations Sustainable Energy For All (SEFA) initiative adopted by all SADC countries.
- The forecasts do not take into account significant economic growth and structural changes in the economic situation of the different countries (see Chapter "The Future of SADC: An Investigation into the Non-political Drivers of Change and Regional Integration"). This is where inter-sector coordination (the nexus) is critical because economic and social activities have a direct correlation to energy demand. The composition of GDP affects energy demand because of the differences in energy intensity which is a measure of the relationship between electricity consumption and economic output. Household consumption is estimated on the basis of the percentage with access disaggregated by social class and the average consumption by each class.

It is also interesting to see the projected electricity demands against the actual sales in each SAPP member country in 2013. This is shown in Fig. 2. Note that in some cases the demand is projected to increase significantly in the short to medium

Table 4 SAPP power and energy demand forecasts

Country	Power (MW)			Energy (GWh)		
	2015	2020	2025	2015	2020	2025
Angola	1657	2226	2871	9437	12,674	16,345
Botswana	711	924	1111	5298	6848	7336
DRC	1510	1772	2054	11,514	13,848	16,915
Lesotho	141	175	215	706	866	1063
Malawi	340	541	629	2347	2833	3293
Mozambique	714	876	1240	4898	5966	7262
Namibia	598	771	830	3956	4838	5767
South Africa	40,659	46,759	54,264	311,474	341,021	365,152
Swaziland	271	304	323	1534	1720	1828
Tanzania	1365	2881	4017	5911	7252	8900
Zambia	2922	3552	4052	15,188	16,168	17,291
Zimbabwe	2535	3174	3751	15,137	18,055	21,295
SAPP Total	53,423	63,955	75,358	387,580	432,089	472,447
SAPP Interconnected	50,061	58,307	67,842	369,885	409,330	443,909

Source SAPP (Unpublished internal report, June 2014)

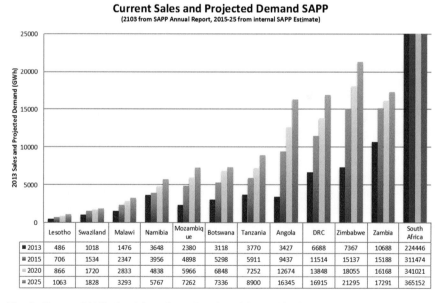

Fig. 2 Current (2013) electricity sales and projected demand in the SAPP

term (2103–2015) from existing sales (e.g. Angola, Zimbabwe). Existing sales are potentially constrained by supply. Demand projections should be closely examined by SAPP to ensure consistency in assumptions.

8   Criteria for Selection of Supply Options

Electricity planners aim to ensure that supply options are reliable (able to meet the expected quality of supply), secure (resilient, able to cope with variability and uncertainty due to human or natural causes) and affordable (available at prices that people are able and willing to pay). In practical terms reliability is achieved by ensuring that the dependable or firm capacity is equal to expected demand (see Text Box 3), security is achieved by having adequate reserve capacity or emergency response measures and affordability is achieved by selecting the least-cost projects, system optimisation and transmission efficiency. In addition to these technical and economic issues, policy and decision makers also consider social and political factors which often determine the projects that get to be implemented.

> **Text Box 3: Sovereign Security in Hydropower Dependent Countries—Firm Versus Average Power and Climate Change**
> Countries planning energy security through hydropower (in the SADC case, primarily Zambia and Malawi), should typically do so by assessing firm power based on the low sequences of hydrological inflows. A climate resilience focus would suggest that this should, as far as practical include some assessment of the potential impact of climate change on runoff.
>
> This would mean that, for the majority of the time, countries should have excess power to sell. Given that hydropower is generally cheaper than thermal generation, and would hold a lower carbon footprint, there would be immediate cost and carbon benefits to thermally dependent countries purchasing that power.

At a regional level SAPP has developed a masterplan in 2009 that demonstrates the reliability and cost savings to be achieved through least cost regional projects, rather than through least-cost national plans. At that time, the least cost regional plan to meet the forecast demand to 2025 was estimated to cost US$89.3 billion while the alternative of meeting demand based on uncoordinated national plans cost US$138.6 billion, a saving of US$47.5 billion. The regional plan had projects with a total capacity of 56,686 MW comprising 18,045 MW hydropower (32 %), 23,883 MW coal (42 %), and 14,758 MW (26 %) gas and petroleum products. The plan was very much dependant on South Africa substituting nuclear power with imports from regional hydropower projects. This was going to require the development of an extra high voltage grid of 765 kV (in contrast to the current levels of 400 kV) from the DRC through the Western and Central Transmission Corridors.

However, in 2010 South Africa developed its own Integrated Resources Plan (IRP) for the period 2010 to 2030 which requires new capacity of 42,600 MW to be met through 6300 MW coal, 9600 MW nuclear, 2600 MW hydropower, 9400 MW solar, 8400 MW wind and 6300 MW gas turbines. In 2012, the SAPP then compiled a list of regional priority projects that take account of the IRP. The list comprises 14,646 MW hydropower (26 %), 9650 MW coal (17 %), 9600 MW nuclear (17 %), 14,100 MW wind and solar (26 %) and 7620 MW gas, heavy and light fuel turbines (14 %). This list of project totals 55,616 MW which is comparable to the 2009 regional plan although with a very different generation mix, with South Africa largely pursuing sovereign electricity security in generation. This would not only require substantially more capital investment, but also would limit (but not eliminate) potential benefits to be had as outlined in Text Box 3. These three plans are compared in Table 5.

The differences in the generation mix is due to the fact that the regional masterplan was optimised on the basis of minimising financial cost whereas the South African IRP, which really determines the regional plan, is based on a multi-criteria approach that takes account of the need to minimise carbon emissions, maximise local employment creation and other economic and social benefits, mitigates uncertainties associated with renewable energy technologies, and minimises water usage. Although support for regional development is considered in the IRP its weighting is less than the other factors.

Coal remains a predominant supply option in South Africa with substantial resource base. The water consumption of this electricity supply option is relatively low but the CO_2 emissions are high.

South Africa in the IRP considers a mix of low carbon supply options including substantial proportions of solar and wind energy. The take up of these options will depend on the investment environment and the incentives for renewables. These supply options also have low water consumption per kWh. Integration to the grid and the "firming" of this temporally variable supply option will be a factor. This issue is discussed further in the section on potential role of hydropower.

Controversy and public perceptions around nuclear options will need to be overcome. However, it is likely that nuclear power will become a more common option internationally as safety matters are resolved and reduction in CO_2 emissions becomes critical.

Gas fields in the North of Mozambique and South eastern Tanzania could provide considerable resource for generation with lower CO_2 emission and low water use per

Table 5 Generation options under consideration		Generation Options (MW)		
		SAPP masterplan	SA IRP	SAPP reg priorities
		2009	2010	2012
	Coal	23,883	6300	9650
	Nuclear		9600	9600
	Hydro	18,045	2600	14,646
	Solar		9400	14,100
	Wind		8400	
	Gas	14,748	6300	7620
		56,676	42,600	55,616

kWh. There are also reserves of shale gas that are proposed for exploitation. The potential impact on water use associated with the "fracking" process (quality and quantity) will need to be considered should this option be explored.

9   Distribution of Resources

It is noted in Chapter "The Future of SADC: An Investigation into the Non-political Drivers of Change and Regional Integration":

> SADC countries with limited hydrocarbon reserves continue to rely heavily on hydro-power, with major facilities planned or under development at sites such as Batoka Gorge (Zambia/Zimbabwe), Lower Kafue Gorge (Zambia), Mphanda Nkuwa (Mozambique) and Inga III (the DRC). Extensions to existing hydropower stations are also envisaged, substantially increasing the present electricity generation portfolio of SADC as a whole.

Table 6 shows the firm and average generation on the Zambezi Basin for Malawi and Zambia. The 2025 demand in both cases is not met by firm hydropower supply, although joint coordinated operation of hydropower means that the firm hydro-power generation is closer to the projected demand (World Bank 2010). Malawi

Table 6 Projected hydropower supply and demand—Malawi and Zambia

Scheme	Firm energy (GWh)	Average (GWh)	Projected demand 2025
Malawi			
Songwe I	21	41	
Songwe II	138	245	
Songwe III	114	207	
Lower Fufu	134	645	
Kholombizo	344	1626	
Nkula Falls	460	1017	
Tedzani	299	721	
Kapichira	541	1063	
Total (GWh)	2051	5565	3293
With coordinated generation	2604		
Zambia			
Batoka Gorge north	950	4800	
Kariba North	3180	4180	
Itezhi Tezhi	284	716	
Kafue Gorge upper	4542	6766	
Kafue Gorge lower	2301	4092	
Total (GWh)	11,257	20,554	17,291
With coordinated generation	14,300		

and Zambia will therefore most likely have to import power at times (unless other electricity sources can be found). However, in both countries the average projected energy production is in excess of 2025 demand, which means that these countries could (if development occurs as planned) opportunistically sell electricity into the SAPP. Given the hydrology of these countries, this could perhaps be done with a lead time of some months—ready to meet winter peak power demands in the 'thermal south'.

10 Hydropower in the Generation Mix

The role of hydropower in the mix of energy options for SADC needs careful consideration of the following:

- The apparently high water 'consumptive use' of this energy option in some project locations (evaporation losses—1000 l/kWh for Kariba versus 1.5 l/kWh for South African thermoelectric[1]) must be weighed against the net economic benefits including the lower carbon footprint. The hydropower water use per unit of electricity will decline as more capacity is installed on the existing dams allowing better usage of the same water. Water use is also lower in the new hydropower stations with smaller reservoir areas.
- The need for a multi-sector approach to optimise the benefits of the water infrastructure according to the needs and priorities of Member States (the water, food and energy nexus). Future national priorities and a desire for sovereign energy, water and food security may preclude cross border electricity trade unless this is seen to be mutually beneficial.
- The inter-annual and seasonal hydrological variability in some basins means the firm energy available from these schemes is a lot lower than the average energy production [e.g. Zambezi hydropower firm = 30,000 GWh versus Average of 55,000 GWh (World Bank 2010)].
- Firm and average energy can be substantially enhanced through joint operation of clusters of schemes (World Bank 2010).
- The World Bank in its multi-sector investment analysis on the Zambezi assumes a value of firm power at US$0.058/kWh, and secondary power at US $0.021/kWh, for its economic analysis. Eskom's estimated costs of generation from coal are US$0.042/kWh[2] (Ham 2012).
- Climate change is projected to significantly reduce the firm and average production from some schemes. While the presence of the dams and associated storage can assist in flood management and water security, the economics of

---

[1]This excludes that water used for coal production, as well as evaporation from storage used primarily for securing water supply to Eskom.

[2]Using 1 US $ = 10 ZAR, and does not include capital redemption on new plants.

current and future projects will need re-consideration (Bleifuss 2012). This also points to an alternative role for hydropower in the regional electricity supply in the medium to longer term.

- The need for substantial upgrades to the transmission network to properly facilitate useful cross border trade and system optimisation. As detailed above, there are several constraints both physical and geo-political, in the construction of these interconnections that need to be overcome.
- How will the Virtual Water transfers be taken into consideration in the future plans for energy security and electricity supply?

In the longer term, once immediate electricity supply constraints have been relieved the more strategic use of hydropower in the overall SAPP system (less base load—more peaking) needs consideration. This would aim to shift hydropower from a base load low value supply option to one where it works in conjunction with other renewable energy options to firm the capacity of solar or wind energy.

11   Climate Change, Virtual Water, Carbon and Cost Benefits of Electricity Trades

A general overview of consumption of water in the generation of electricity and the consequent Virtual Water embedded in electricity trades is provided in Chapter "Quantifying Virtual Water Flows in the 12 Continental Countries of SADC". However, such a broad assessment only provides a limited overview of the potential for water, carbon and cost benefits to be had in electricity trades through the SAPP, and detailed assessments of specific transfers will need to be considered. The import and export of electricity across the SAPP depends on the supply and demand situation in any particular year, while the actual trades may be constrained by a lack of suitable transmission infrastructure. The future electricity supply situation in turn depends on the construction of the planned infrastructure (generation and transmission), the maintenance requirements of generation infrastructure, and at least in the case of hydropower the rainfall and runoff. Demands will depend on economic growth, the nature of that growth as well as short term weather events (in South Africa, particularly cold weather in the winter is causing peak demands).

As outlined in previous chapters, the impacts of climate change on the electricity supply situation in SADC as a whole might be heavily dependent on the position of the Inter-tropical convergence zone (ITCZ) across the Zambezi Basin in that year. Climate projections generally suggest that the area north of the ITCZ will receive more rain, while that area to the south may get less. The enormous complexity of the meteorological conditions determining the position of the ITCZ makes it difficult to postulate its likely movement under climate change, and global climate models often differ in their predictions. The additional complexity of rainfall/runoff ratios makes it even more difficult to project the impacts on hydropower in the Zambezi. Nonetheless, both World Bank and InternationalRivers studies suggest up to a 50 % decrease in firm power, and an average power decrease of some 25 %. In spite of this Table 6 shows that *on average*, Zambia and Malawi *could potentially* have an average surplus of 5535 GWh/a, should all the planned hydropower be installed.

This *average* surplus, if sold into the SAPP, has Virtual Water, carbon and cost benefits to the region and particularly South Africa. Sold into the SAPP at the secondary power value used in the World Bank study, it has a value of some US \$116 million (as an annual average). The cost of generating that power in South Africa's coal fired thermal stations is twice that value. The water saved, using the average water use in ESKOM's thermal stations would be nearly 8 million m^3/a, which is 2.5 % of that facility's total water use. This total is enough to irrigate about 800 ha or provide free basic water (at 6000 l/household/month) to over 100,000 houses for a year. ESKOM's CO_2 emission factor is 0.99 kg/kWh (Urban Earth 2012) for 2012, with a target of 0.68 kg/kWh. This means a CO_2 emission saving in the average surplus (assuming zero emissions from hydropower) of some 5500 million tonnes or some 2.4 % of the facility's total annual emissions. This compares favourably with Kyoto's 8 % emission reduction target although ESKOM's other initiatives could reduce total emissions by more than 30 %.

While these analyses are, given the data available, necessarily superficial they do nevertheless show that there is some potential for water, carbon and cost savings in electricity trading through the SAPP, which would warrant further investigation.

12 Conclusions and Recommendations

SAPP uses national plans as input but has no mandate to influence national planning except by providing non-binding regional masterplans. Unfortunately, national plans are developed without taking account of regional opportunities resulting in sub-optimal developments, but with sovereign electricity security advantages. Clearly political and social factors can, and do, often override technical and economic considerations in the energy sector.

While it is easy to demonstrate the benefits of regional cooperation in economic, water and carbon terms, introducing this into national planning processes will be challenging. It is therefore important that the core message carried to national policy and decision makers shows the relative advantages of regional integration, after electricity security and other legitimate national concerns have been addressed.

Perhaps more importantly, these messages must be based on sound analysis and the best available information. The experiences in the SAPP highlight that regional cooperation requires the adoption of multi-sector planning approach to establish credible demand forecasts and multi-criteria approaches to selection of project options that go beyond identification of the least cost to incorporate the national interests of participating countries. This can be achieved by using an iterative planning process as follows:

- Harmonisation of planning criteria and minimising information and data asymmetries—policy and decision makers need to agree on a common planning horizon and to invest resources to establish demand forecasts and project feasibility studies for candidate projects to be considered for national and regional plans.
- Development of national plans that take account of non-negotiable and negotiable national interests.
- Development of regional plans that respect national interests of participating countries but provide information on the economic and other benefits of regional cooperation.
- Review of national plans to maximise opportunities for reliability and economy afforded through regional cooperation. Countries can also make informed cost benefit analyses of the additional costs for national security interests which would now be explicit.

This process should also be based on:

- An improved understanding of the impacts of climate change on firm and average power, and the availability of surplus hydropower for sale through improved downscaling of general circulation models (GCMs).
- Identification of methods to provide 'early warning' of the likelihood of surplus power, and when that may be made available on the annual and daily cycle, for example it has been shown that the flow in the Zambezi is heavily reliant on the previous season's rainfall.
- An assessment of the feasibility and operation of systems to 'ramp down' thermal power plants to maximise opportunities for water, carbon and cost savings—the need for maintenance and margins. This would include the impacts on coal purchase contracts, and the opportunities to sell the coal to other buyers.

It is believed that this iterative process will lead to win-win outcomes in which each country will have internal generating capacity that can be used to meet local demand in emergency situations but also having the opportunity to benefit by purchasing power from the least-cost regional sources under normal operating conditions. The development of large hydropower in the region can help in the development of a clean low carbon energy future by supporting the deployment of more solar and wind energy, and using the considerable average energy potential to reduce operating costs for thermal systems.

References

Bleifuss J (2012) A risky climate for southern African Hydro: assessing the hydrological risks and consequences for Zambezi Basin Dams. International Rivers. Available from http://www.internationalrivers.org/files/attached-files/zambezi_climate_report_final.pdf. Accessed 9 July 2015

Ham J (2012) Knowledge piece published on Probus. http://www.probus.org.za/cost-of-power.
    html. Accessed 10 July 2015
Southern African Power Pool (2013) Annual report 2013
Urban Earth (2012) Eskom announces unchanged electricity emission factor for 2012. http://
    urbanearth.co.za/articles/eskom-announces-unchanged-electricity-emission-factor-2012.
    Accessed 29 June 2015
World Bank (2010) The Zambezi basin: a multi-sector investment opportunities analysis. http://
    siteresources.worldbank.org/INTAFRICA/Resources/Zambezi_MSIOA_-_Vol_1_-_
    Summary_Report.pdf. Accessed 5 July 2015

Virtual Water and the Private Sector

Guy Pegram and Hannah Baleta

Abstract Water is strategically important in the Southern African Development Community (SADC) for the production of agricultural products, industry, mining and power generation. Water is also important in food security and poverty eradication through the support of a range of livelihood practises. Water in southern Africa is stressed; therefore unlimited use of the resource is not possible. Trade-offs between one water user and another become important discussions as scarcity progressively increases within the region. There exists an opportunity for the private sector to formulate partnerships within the water sector which promote adequate water resources management and governance. This chapter explores the nature of private sector engagement with managing water risk, and the use of water footprinting as a tool to facilitate this. Already water footprints have been used by the private sector in assessing their business risks with a product supply chain, operations or a region—including the risks posed by climate change. This chapter indicates that the concept of virtual water may also be a useful tool in building dialogue between the public and private sector.

1 Why Is the Private Sector Engaged in Water?

1.1 Water Security Risks Are Becoming Increasingly Important for Businesses

Corporates are increasingly beginning to recognise the risks posed by an insecure water future. In the recent World Economic Forum (WEF) Global Risks Report (2014), water crises were ranked as the third highest global risks of highest concern in 2014. This illustrates the continued and growing awareness of the global water crisis

G. Pegram (✉) · H. Baleta
Pegasys, Cape Town, South Africa
e-mail: guy@pegasys.co.za

H. Baleta
e-mail: hannah@pegasys.co.za

© Springer International Publishing Switzerland 2016
A. Entholzner and C. Reeve (eds.), *Building Climate Resilience through Virtual Water and Nexus Thinking in the Southern African Development Community*, Springer Water, DOI 10.1007/978-3-319-28464-4_7

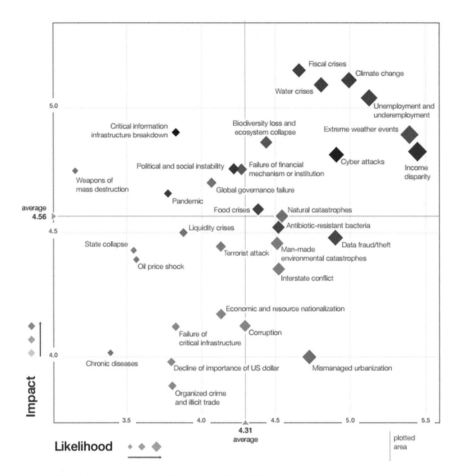

Fig. 1 Global risks landscape, 2014. *Source* WEF (2014)

as a result of mismanagement and competition of scarce water resources. Considered together with other high ranking risks (Fig. 1) such as climate change, extreme weather events, biodiversity loss, environmental concerns and income disparity, water security throughout the value chain and with respect to their 'social licence to operate' are becoming increasingly prominent in corporate decision making.

Not only is water quantity a concern, but water quality too. Pollution incidents have been known to paralyse business operations in parts of China and elsewhere on the globe, while the impacts of corporates on the quality of domestic water supply has come under increasing scrutiny. As competition for the scarce resource increases, the private sector needs to consistently show their commitment to sustainable water resources management.

The increasing awareness of water security risks, as indicated by the WEF Global Risks Report (2014), has led to increasing interest from the private sector in mitigating their water security concerns. Recognising that these risks cannot be mitigated alone, there has been growing interest in collaboration and partnership in order to manage water risks within a catchment.

1.2 Corporate Experiences of Water Security Risks

The risks highlighted by WEF are not purely academic in nature. There are a number of cases globally where a significant physical, regulatory or reputational risk was experienced by companies due to water security risks within a catchment. Poor water quality or water shortages are often blamed on business, even when businesses fully comply with regulatory requirements. Regardless of whether the private sector is solely responsible or not, they are often a major economic contributor within a catchment. Therefore, the perception of being a water polluter for example may be as damaging.

1.2.1 Lake Naivasha Cut Flowers (WWF 2012)

Situated in the Rift Valley of Kenya, Lake Naivasha is an internationally renowned Ramsar site. The Lake is also home to a blossoming agricultural industry, exporting high value fresh vegetables and cut-flowers to European and English markets. Popular news implicates the lake's ecological degradation on the cut-flower industry surrounding the lake. As a result, the valuable GDP contribution of the flower industry to Kenya has been at risk, as retailers come under scrutiny of consumers in Europe and England. With the onset of a drought in 2009, tensions within the basin began to escalate as competing water users such as agriculture, livelihoods and the functioning lake ecosystem became under pressure.

The response to increasing pressure on Lake Naivasha between all of the competing users was to collectively optimise the management of the water resource. Increasing water use efficiency was carried out through three interlocking strategies: improved governance, fostering partnerships and promoting more responsible individual water use. With solid governance, regulation and enforcement, a broad framework is created which is able to incentivise water users to be more responsible. In the context of Naivasha, partnerships between the different water users helped through the sharing of resources, skills and knowledge, building institutional capacity and collaboration. Finally, the benefits of individual water users taking responsibility of their actions and pursuing better practises that are attuned to the local social and hydrological context (WWF 2012).

Future irrigation or urban abstraction will continue in Lake Naivasha, further pressurising the system in parallel with climate change. The already significant development impacts will continue and increase over time due to development pressures, population increase and economic growth in the country. Without adequate communication and collaboration between the stakeholders in the basin, social and economic development will not take place alongside an ecologically sustainable Lake.

1.2.2 Kerala Coca-Cola (the Rights to Water and Sanitation, n.d.)

In 2004, Coca-Cola experienced intense reputational risk as a bottling plant in South Kerala, India was forced to close down after residents claimed it was draining and polluting their ground water supplies. Globally, the reputational brand value of Coca-Cola was believed to decrease in the eyes of many consumers. The incident took place during a time of drought. One school of thought believes that the meteorological effects of the drought would have led to equivalent water shortages in the region, regardless of the water use of Coca-Cola.

The incident in India galvanised awareness about water risk among global companies, catalysing action such as collective action, in partnership with communities or public sector organisations in basins which face particular water risks. The incident forced the recognition that perceived negative impacts to the local water supply are as dangerous as actual impacts, due to the civil action and loss of social licence to operate for Coca-Cola.

As a result of experiences such as in Kerala, Coca-Cola has a well-developed strategy for investigating their water risks associated with particular bottling facilities across the globe.

1.2.3 Peru Asparagus (Progressio 2010)

A high level search on news regarding Peru asparagus brings the following headline from The Guardian (Lawrence 2010): "How Peru's wells are being sucked dry by British love of asparagus." This headline indicates the risks associated with production of particular high-value crops (such as Asparagus) in areas which are water scarce (such as the Ica Valley in Peru). This is especially the case when the high value crops are not consumed within the country, but instead exported to Europe, England or the USA. In some cases in the Ica Valley of Peru, the aquifers are dropping by 8 m year, the fastest rate of aquifer depletion in the world.

The complexity arises when considering the economic value of the crop export to the economy of Peru, where the Ica Valley is home to 95 % of Peru's asparagus. The awareness of consumers in the UK and EU has led to some innovative measures in order to reduce the significant water use of the agricultural practise. Water footprinting has been instrumental in bringing this subject to light. The trade of produce with high amounts of embedded water has been initiated through the concept of virtual water and water footprinting.

1.3   Investor Concerns

Not only are companies, but many investors are also becoming increasingly concerned with respect to water security risk and their investments globally. This is
evident through the increase in tools and guidance such as the Lloyd's 360 risk
report on global water scarcity (Lloyd's 2010) or the Carbon Disclosure Project
(CDP) water report (CDP 2015).

"In Deloitte's work with the Carbon Disclosure Project to survey and report on
executive awareness of global water issues, we have seen an upward trend over the
past couple of years in the perceived level of risk exposure associated with water
scarcity in both direct operations and supply chain operations. In our 2011 survey,
we saw 59 % reporting at least one such risk, and more than 65 % of these same
respondents report having a potential for this type negative impact now or within
five years." (Deloitte 2012)

Water insecurity may also be seen as an opportunity. In the CDP report (2012),
63 % of respondents consider water scarcity as a potential driver of innovation in
their companies. Opportunities in water efficiency, revenue from new water products or services, and improved brand value through effective handling of water
issues are some of the opportunities listed.

The interest of investors is pertinent due to the potential disclosure requirements
which may be needed of companies should they begin stakeholder engagement to
mitigate water risks for example. A lack of disclosure, implying a lack of adequate
consideration of the challenge, may forfeit particular sources of funding for projects
which are particularly at risk.

1.4   Motivations for Engagement

The motivations for engagement may be varied from the private sector perspective.
Water risk may result in a range of associated risks to the private sector, as indicated
in Fig. 2. Risks include physical risk, which is directly related to too little (scarcity),

Fig. 2 Managing the suite of
water risks. *Source* Pegram
et al. (2009)

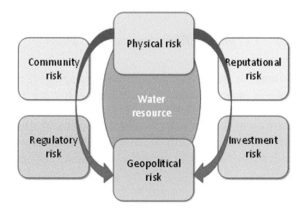

too much water (flooding) or water this is unfit for use (pollution), as well as water quality risks (not necessarily directly associated with the business operation), each of which is associated with the management of the water resources. Regulatory risks are related to the government's management of water resources, particularly during times of physical water risk. Reputational risk is related to the exposure of companies to customer purchasing decisions, associated with the perceptions around business decisions, actions or impacts on water resources, aquatic ecosystems and communities that depend on them (Pegram et al. 2009).

The previous three examples (Lake Naivasha, Kerala and Peru), although resulting in different forms of engagement around water risk mitigation, all stemmed from a reactive response to an operational crisis. In all three cases, the future of the water supply was under threat.

A range of motivations for engagement from the private sector include the following:

- A reactive response to existing operational crisis as seen in the example from Kerala, India. Following that, Coca-Cola has changed the procedure in which risks are analyzed within catchments where they operate.
- A strategic risk in the future to operations or supply chains is often carried out once the current risk to the operation or supply chain is explored.
- External pressure from investors and consumer advocates is evident through a number of requirements from disclosure reports such as CDP Water or corporate governance requirements.
- Leadership positioning related to corporate social responsibility may be an additional driver for engaging water stewardship initiatives to minimize water risk.
- Competitive advantage in marketing the company may be an additional benefit to managing water risks in a collective manner within a catchment. It is critical that this is done strategically alongside other drivers for engagement to reduce the perception of "green washing."

As indicated, there are a range of drivers motivating engagement of the private sector in water risk management. These drivers are not mutually exclusive, and can be used as a suite of motivating factors for a company to engage.

1.5 Corporate Risk Assessment Tools

In order to improve corporate water management, and ultimately to support sustainable water resources management, corporates need to be cognisant of their particular water use and related impacts. Without knowing their relative total water use footprint or polluting effect, real engagement cannot take place. Corporate water accounting needs to be done in unity with other corporates and public institutions to understand the broader context of water risks within a catchment (Fig. 3).

Fig. 3 A range of water risk assessment tools. *Source* DEG-WWF (2015), WBCSD (2015), ISO (2014), WFN (2015), Aqueduct, n.d. and Ceres, n.d.

There are a wide range of corporate water risk assessment tools which have been developed to support companies or investors in assessing the water risks being faced. Depending on the audience or the nature of the risk, different tools may be used. This may also pose a risk in some cases, as corporates have been known to use assessment tools only when the results suit their needs. The challenge of business interests driving the metrics used in the water sector is a real concern. Critical analysis needs to take place before taking on board assessment tools, to ensure that the metric has not been developed solely to support a business's own interests. Therefore there are challenges associated with which risk assessment tool or guide to follow because the development of corporate interest is relatively new.

Tools may either be web-based or excel based, and may use global data-sets or be based on internal facility information alone. Some of the most often-used tools include the DEG-WWF Water Risk Filter (DEG-WWF 2015), Aqueduct Water Risk Maps (Aqueduct, n.d.), Ceres Aqua Gauge (Ceres, n.d.) Water Tool and the WBCSD Water Tool (WBCSD 2015). The ISO have also been developing a water risk assessment guideline, and have been moving towards the establishment of an ISO standard for water footprinting (ISO 2014).

Water footprinting is one of these tools, which has been used by a range of companies in assessing the intensity of their water use in relation to a wider catchment. Water footprinting was originally developed as a quantification of the concept of Virtual Water (Allan 2011). It has since been used in the calculation of a number of trade-flows, indicating the movement of embedded water between countries.

The Water Footprint Network (WFN 2015) has done major work in establishing guidelines on how to carry out the water footprint methodology at a number of scales, including a facility, product, regional or national level. Corporate water footprints measure the total volume of water used directly and indirectly to run and support a business. The corporate water footprint is equal to the sum of the direct and indirect water footprints of all products including raw material production, manufacturing and distribution.

2 Where Does the Water Footprint Have Traction?

As indicated previously, some water risk assessments are useful depending on the information availability or risks of concern. The same is true for water footprinting. The concept of the water footprint has traction in a number of ways for the private sector. The areas of traction include understanding the water footprint of: supply chains, operations, use phase and regional implications. These areas of traction are supported by high level case examples.

2.1 Supply Chain

At a supply chain level, the concept of the water footprint is especially useful to identify high water intensive inputs into your product and the source of that water (as blue water—extracted from surface or ground water or green water derived from rainfall and soil moisture). This can also be done for inputs across a range of regions, to identify the water risks associated with sourcing inputs from one area as opposed to another. Chapter "Mechanisms to Influence Water Allocations on a Regional or National Basis" on Water Allocation highlights the role that this may play in building climate resilience.

2.1.1 SAB Miller (SABMiller, n.d.)

SAB Miller is one of the most well-known cases of a company carrying out a water footprint for the purpose of supply chain water intensity identification. SABMiller carried out a water footprint for their supply chain, calculating the water embedded into their entire value chain, across a range of locations where they source their inputs (Fig. 4). They found that the majority of the embedded water in their beer products was situated in the cultivation of the crops used in beer (i.e. barley and maize represent 90 %). The relative water use in the crop processing, brewing, distribution and consumer phase of the product lifecycle was significantly less (10 %).

In addition to finding of the relative intensity of embedded water across the supply chain, SAB Miller also investigated the relative water footprint of crops such as barley, or hops or maize across a number of countries globally.

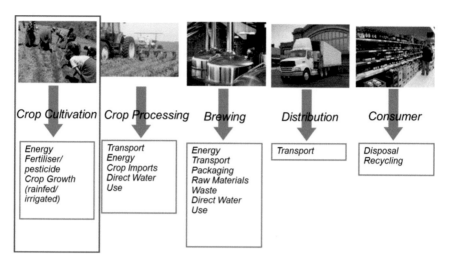

Fig. 4 Beer production supply chain. *Source* SABMiller, n.d.

A comparison was done between beers produced in South Africa vs. the Czech Republic. It was found that, in absolute numbers, the water footprint for beer in South Africa (155 l/l of beer) was more than three times the Czech Republic's footprint (45 l/l of beer). The SAB Miller Report is quick to add that this is not due to any difference in efficiencies, but rather due to the higher evaporative demand for crops in South Africa and the increased reliance on irrigation, adding a significant blue water portion. Nonetheless, that in itself is a telling metric with respect to climate change.

Therefore, a supply chain water footprint can be helpful to corporates, not only in identifying the relative embedded water requirements across the supply chain, but also to compare supply chains across different regions.

2.1.2 Illovo (Zambia Sugar)

Like SAB Miller, Illovo have carried out a water footprint for their supply chain. Focussing on their major operations in Zambia Sugar, they focussed on the nature of the different water types, and therefore the relative requirements of each in terms of quality and assurance of supply. The majority of their water footprint stems from crop cultivation, with a large proportion attributed to blue water irrigation. The value of doing the water footprint, was to better understand the relative water use in the context of an already water stressed Kafue River Basin in Zambia.

Setting these findings against the water footprint of total agricultural exports from Zambia also provides for useful metrics. CRIDF's Virtual Water database shows that sugar exports from Zambia include some 72 % blue water and some 28 % green water, and earn the country some U$128 million/a in export earnings. Total agricultural exports from Zambia include some 8 % blue water, and 88 %

green water, earning some U$ 880 million in export earnings. Illustrating the relative importance of irrigation of sugar to Zambia's agricultural export earnings, and hence its vulnerability to the potential combined effects of reduced rainfall and runoff, as well as competition with other users.

2.2   Operations

Some corporates may use the water footprint methodology at the operational level. In this case, the water footprint alone may not be as useful, as the number itself may vary widely depending on the growing season, variability of rainfall or temperature or even where the crop is grown. The tool is useful however, in conjunction with additional information collected at an operational level.

2.2.1   Coca Cola (CocaCola and the Nature Conservancy 2010)

Coca-Cola uses a range of tools in the determination of their operations facility vulnerability to water risk. The company has faced significant challenges in the past with regards to water security. In many cases, it was a lack of knowledge regarding the wider catchment water security and social context which led to the significant reputational risks in Kerala for example.

In light of the risks which Coca-Cola has faced, they use systemic method to identifying and investigating all of the risks in a catchment. Through a tiered approach, they have a framework which considers where the plant is located, the relative water use of the facility in the catchment and other indicators which may give light to the water risks in the catchment. These elements include the runoff, irrigation, night lights and population of the region where the plant is located. The wide range of information helps identify not only physical, but reputational and regulatory risks associated with the local social and economic context.

A water footprint alone to identify and motivate corporate water risks at an operational level is not sufficient. However, as indicated by Coca-Cola, when supported with additional information, a water footprint may be useful.

2.3   Use Phase

Traditionally, most water footprinting analyses consider the upstream water use of a product of facility in order to understand the relative weight of embedded water in the product or facility. In some cases however, a product may have a higher water footprint at the consumer use phase. This is the case for P&G Fabric Care.

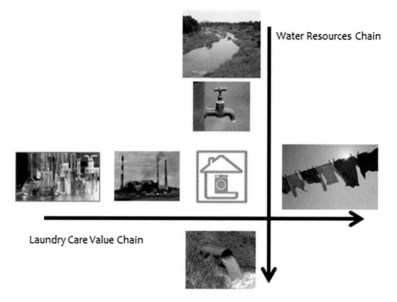

Fig. 5 Laundry care value chain and water resources chain

2.3.1 P&G Fabric Care

P&G Fabric Care have carried out a high level water footprint analysis to identify the relative importance of efficient water use at the consumer level for their product. Unsurprisingly, the majority of embedded water in fabric products is at the user stage. This has implications for the focus of P&G in terms of being good water stewards. In light of the information from the water footprinting assessment, their focus may arguably need to be around reducing the rinse requirements of their products or supporting water saving techniques at household level (Fig. 5).

The value chain of fabric care products is essentially the complete opposite to that of beer production for SABMiller, where 90 % of the embedded water is attributed to the consumer use phase. A water footprint is useful in this regard for corporates due to the ability to clearly communicate to consumers the relative importance of their washing practises for water efficiency. The water footprint is also useful in this respect in justifying additional focus from P&G in managing their future customers water needs to secure the future market.

2.4 Regional Implications

Water footprints need not be done on a single product from a corporate perspective. Corporates may also compare, or analyse their respective water footprints and associated risks across a region. Carrying out a water footprint in relation to the water scarcity context in which the footprint is situated is particularly helpful in identifying water risks both for the company, but the catchment in which they are situated too.

2.4.1 SAB Miller (SABMiller, n.d.)

SAB Miller has carried out a range of water footprints for their products across the globe. As indicated previously, the water footprint to produce a litre of beer in South Africa is three times that of The Czech Republic.

SAB Miller have further taken the water footprint of their respective input needs, and overlaid this with an understanding of water scarcity in the region where produced. Although the water footprint for beer in South Africa is higher, the local context of being a South African company, with inputs being sourced from its home country, makes it difficult to move input sources. Indeed, as Muller's paper (Chapter "Virtual Water and the Nexus in National Development Planning") points out, water is usually a relatively less important metric in the siting of large developments. Nonetheless, the value of the water footprint at a spatial level is the insight of overlaying where inputs are sources with areas of significant water scarcity within South Africa. Through assessments such as these, the information gathered can help substantiate a business case for increased engagement in water saving technology, or catchment management for example (Fig. 6).

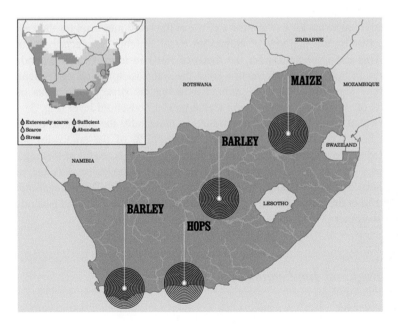

Fig. 6 Crop growing regions across South Africa (*large map*) and annual renewable water supply per person (*insert map*). *Source* SABMiller, n.d.

2.4.2 Illovo Sugar

Illovo Sugar has also carried out an assessment of their product water footprints at a regional scale. In the case of Illovo Sugar, the focus of the water footprint method was to identify the different nature of water use for sugar production in each Southern African country. With an improved understanding of water use needs (i.e. blue vs. green), Illovo are better able to understand their risk to droughts or climate change. Where there is a larger blue water footprint, irrigation is used more, and therefore vulnerability to irregular rainfall is reduced, although in the absence of significant storage the vulnerability to the multiplier effects of reduced runoff coefficients and reduced rainfall increases. In the case of Illovo Sugar, the exact water footprint number is less of relevance than using the concept to map and identify approximate areas of concern.

Figure 7 indicates the Illovo Sugar major sugar cane growing regions. The base map is the WWF-DEG Water Risk Filter, indicated areas which experience water stress. The map shows the vulnerability of sugar cane grown in Swaziland and South Africa to drought stress and competition for water. However, as indicated in the annotations, these areas are mostly irrigated cane growing regions with significant storage, and are therefore protected against some of the climate variability. The increased irrigation however, means that the water use is in competition with other users such as industry or urban development, and therefore may still be under threat.

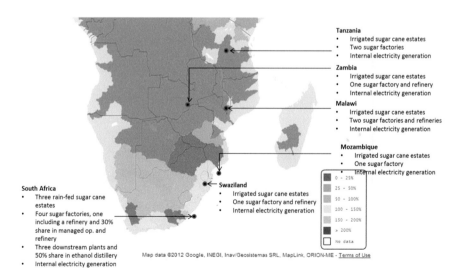

Fig. 7 Illovo mills in southern Africa superimposed on Aqueduct Water Stress map. *Source* Aqueduct (n.d.) and authors own labels

3 What Is the Value of the Water Footprint to the Private and Public Sector?

The concept of the water footprint is useful to the private sector in better understanding their supply chain, operational, consumer and regional embedded water use. However, the concept is also valuable in building dialogue between the private and public sector.

As indicated in Fig. 8, the concept of the water footprint helps to translate 'water-centric' concepts of water resources into more politically tangible 'water-supportive' concepts such as economic activity (GDP contribution or employment) or even further into trade. The development of a common language for both the private sector and public sector to understand the role of water in the economy is invaluable in the creation of conversations around economic investment and development.

At a basic level water footprinting is useful at the each scale for the answering the following key questions:

- Supply Chain

  - Where is water use the most intensive?

This helps corporates or farmers identify how to use water better, ensuring less loss at a field and system level. At a supply chain level, a water footprint may also help in using water smarter and selecting the best crops/practices for the region.

At a consumption orientated supply chain level, the water footprint can help in education towards consuming less embedded water.

- Operations

  - How to use less water better?

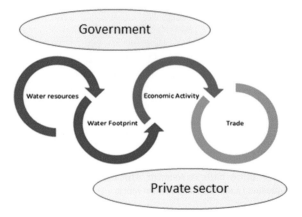

Fig. 8 The role of water footprinting in building dialogue between the private and public sector

Like in the supply chain, at an operational level, the water footprint is useful in identifying where the majority of water is being used, so that efficiencies can be gained.

- Regional level

  - Is this the best use of water considering the stress and economic value of the product?

A water footprint at a regional level may help with the identification of economic activities using comparative advantage. Comparing the relative water use intensity vs. the GDP contribution or employment may support decision-makers at a regional level with regards to which economic activities to support.

The regional water footprint lens may also help identify development opportunities for livelihoods and equity. The uses of a water footprint at a regional scale are also especially helpful for building dialogue. Moreover, for corporates like SABMiller and Illovo there may be opportunities for regional risk pooling to increase climate resilience, particularly in terms of limiting the risks to outgrower schemes.

- Dialogue

  - WFs help to develop a common understanding of the value of water in the economy between government and business.

Effective dialogue between the private and public sector is absolutely necessary for risk sharing and collaborative management of water resources to take place. The concept of embedded water and water footprinting is especially useful in highlighting the water imports and exports associated with trade. This may help in the selection of developing particular products for import or export, or highlighting the contribution of rain-fed or irrigated crop production to the national economy.

The value of understanding the water footprint between different sectors and regions is also through the potential for regional integration. The concept of embedded water is useful in illustrating the concept of benefit sharing between countries.

At a Southern African Development Community (SADC) level, water is stressed. Although water footprinting alone is unable to meet all the needs for adequate water resources management, it is a useful tool for the purposes mentioned above. In many parts of SADC for example, crops are grown where the embedded water is far greater than if the crop was grown elsewhere. Notwithstanding the hydro-centric versus hydro-supportive concepts debated in Chapter "The Future of SADC: An Investigation into the Non-political Drivers of Change and Regional Integration", and the fact that the locations for crop production are influenced by a number of metrics, the concept of embedded water allows planners to compare the comparative advantage of particular crops and economic activities, in light of their water needs.

The real benefit of using the concept of embedded water in the SADC region however, is through the development of a dialogue. Whether between the private and public sectors or between different countries within SADC, the use of a neutral water risk assessment tool will greatly benefit communication. As indicated previously, the tool alone is not adequate. However, it is able to translate complex water resources and economic jargon into accessible illustrations of the flow of water through an economy.

References

Allan T (2011) Virtual water: tackling the threat to our planet's most precious resource. I.B.Tauris & Co Ltd

Aqueduct (n.d.) Water Risk Atlas. http://www.wri.org/applications/maps/aqueduct-atlas. Accessed 2 Oct 2015

Ceres (n.d.) Aqua Gauge. http://www.ceres.org/resources/reports/aqua-gauge/view. Accessed 2 Oct 2015

CDP—Carbon Disclosure Project (2012) Water disclosure global report. https://www.cdp.net/cdpresults/cdp-water-disclosure-global-report-2012.pdf. Accessed 2 Oct 2015

CDP—Carbon Disclosure Project (2015) Annual CPD program reports. https://www.cdp.net/en-US/Results/Pages/reports.aspx. Accessed 26 Sept 2015

Coca Cola and The Nature Conservancy (2010) Product water footprint assessments. http://assets.coca-colacompany.com/6f/61/43df76c8466d97c073675d1c5f65/TCCC_TNC_WaterFootprintAssessments.pdf. Accessed 2 Oct 2015

ISO (2014) Water footprint—Principles, requirements and guidelines. http://www.iso.org/iso/catalogue_detail?csnumber=43263. Accessed 27 Sept 2015

Lawrence F (2010) How Peru's wells are being sucked dry by British love of asparagus. The Guardian, 15 Sept. http://www.theguardian.com/environment/2010/sep/15/peru-asparagus-british-wells. Accessed 26 Sept 2015

Lloyd's (2010) Lloyd's 360 risk insight: global water scarcity—risks and challenges for business, London, UK. http://awsassets.panda.org/downloads/lloyds_global_water_scarcity.pdf. Accessed 26 Sept 2015

Pavlovsky K, Sarni W (2012) Investor interest in water risk on the rise (Deloitte's Sustainable Business Blog, 17 May). http://blogs.deloitte.com/greenbusiness/2012/05/investor-interest-in-water-risk-on-the-rise.html. Accessed 27 Sept 2015

Pegram G, Orr S, Williams C (2009) Investigating shared risk in water: corporate engagement with the public policy process. WWF, Surrey, UK. http://www.pegasys.co.za/pdf_publications/Investigating%20Shared%20Risk%20in%20Water_Corporate%20Engagement%20with%20the%20Public%20Policy%20Process.pdf. Accessed 27 Sept 2015

Progressio (2010) Drop by drop: understanding the impacts of the UK's water footprint through a case study of Peruvian asparagus, London, UK. http://www.progressio.org.uk/sites/default/files/Drop-by-drop_Progressio_Sept-2010.pdf. Accessed 26 Sept 2015

SABMiller (n.d.) Water footprinting: identifying and addressing water risks in the value chain. SABMiller and WWF, Surrey, UK. http://www.sab.co.za/sablimited/action/media/downloadFile?media_fileid=918. Accessed 27 Sept 2015

The Rights to Water and Sanitation (n.d.) Case against Coca-Cola Kerala State: India. http://www.righttowater.info/rights-in-practice/legal-approach-case-studies/case-against-coca-cola-kerala-state-india/. Accessed 2 Oct 2015

Water Footprint Network (2015) Securing fresh water for everyone http://waterfootprint.org/en/. Accessed 27 Sept 2015

WBCSD (2015) Water tool. http://www.wbcsd.org/work-program/sector-projects/water/global-water-tool.aspx. Accessed 27 Sept 2015

WEF—World Economic Forum (2014) Global risks 2014, 9th edn. Geneva, Switzerland. http://www3.weforum.org/docs/WEF_GlobalRisks_Report_2014.pdf. Accessed 23 Sept 2015

WWF—World Wide Fund for Nature (2012) Shared risk and opportunity in water resources: seeking a sustainable future for Lake Naivasha. Gland, Switzerland. http://awsassets.panda.org/downloads/navaisha_final_08_12_lr.pdf. Accessed 26 Sept 2015

WWF-DEG Knowledgebase (2015) Water risk filter. http://waterriskfilter.panda.org. Accessed 27 Sept 2015

The International Experience

John Anthony Allan

Abstract The first purpose of this chapter is to show that there are a number of international conditions and trends, first in water, energy and food supply chains and secondly, in international trade and demography that need to be understood. These conditions and trends are very relevant to understanding future options in the allocation, management and consumption of water and energy in southern Africa. Secondly, it will highlight the recent heightened awareness amongst scientist and resources managing professionals of the ways that over-consumption and poor stewardship of natural resources have impacted the sustainability of three key strategic supply chains—water and sewage services, energy services and food. Thirdly, it will be noted that these supply chains are shaped in major ways by private sector markets and international trade. These markets are badly regulated because they do not have accounting rules that generate market signals that incentivise resource stewardship and sustainable and ethical investment. Water for example is not valued in food supply chains. There are many casualties in these highly distorted systems of provision. Fourthly, because water, energy and food are vital and therefore very easily politicised all three supply chains have been impacted and distorted in major ways by government subsidies and related payments. In these politicised contexts there are some very powerful players. They are powerful because they are deeply informed, they have significant contractual and commercial leverage and they operate at a scale that can influence public policy as well be responsive to it. The asymmetric power relations make sustainable practices impossible for the weak supply chain players. These sub-optimal systems characterise the water, energy and food nexus world-wide. An additional reason for analysing the water, energy and food nexus from a supply chain perspective is that those who operate these market and public systems are deeply informed on what is politically feasible and can identify the reforms that are possible as well as needed.

J.A. Allan (✉)
Department of Geography, King's College, London, UK
e-mail: ta1@soas.ac.uk

© Springer International Publishing Switzerland 2016
A. Entholzner and C. Reeve (eds.), *Building Climate Resilience through Virtual Water and Nexus Thinking in the Southern African Development Community*, Springer Water, DOI 10.1007/978-3-319-28464-4_8

1   International Contexts and Trends in Water, Energy and Food Consumption

The first purpose of this Chapter is to show that there are a number of international conditions and trends, first in water, energy and food supply chains and secondly, in international trade and global demography that need to be better understood. These conditions and trends explain the way natural resources are consumed and stewarded in current supply chains as discussed in other Chapters of this volume. They are also very relevant to understanding future options in the allocation, management and consumption of water and energy in southern Africa.

Secondly, it will highlight the heightened awareness amongst scientist and resources managing professionals of the ways that the over-consumption and poor stewardship of natural resources internationally impact the sustainability of three key strategic supply chains (1) water and sewage services (2) energy services, and (3) food provision. Those managing these separately operated supply chains have aimed to optimise the consumption of either water or energy—rather than both, in their respective supply chains. They have presumed that the way they operate is efficient because they have complied with accounting rules put in place by their national accounting standards boards. But these accounting rules are in practice very dangerous because they ignore the costs of water as an input and underestimate the costs of energy. In addition, they do not capture the costs (externalities) of mismanaging water and energy.

Until recently those providing water and energy services and food have ignored the risks of operating in water scarce locations. They have also ignored the effects on ecosystems of operating in ways that negatively impact them. Nor have the risks of producing energy in ways that negatively impact water resource availability and the health of water ecosystems. Awareness of these complex impacts led to the development of the concept of the nexus which highlights the need to manage water and energy in ways that are sustainable in the long term.

Thirdly, it will be noted that these supply chains are shaped in major ways by private sector markets and by international trading transactions. There are many casualties in these sub-optimal national and international systems of provision. For example farm livelihoods are impoverished by low food prices and farmers cannot afford to provide essential ecosystem stewardship services.

Fourthly, it will show that because water, energy and food are all vital and therefore very easily politicised all three supply chains have been impacted and distorted in major ways by government subsidies and related payments. In these politicised contexts there are some powerful players that can influence international transactions. They are powerful because these players are very deeply informed, they have significant contractual and commercial leverage and they operate at a scale that can influence public policy as well be responsive to it. The asymmetric power relations make sustainable practices and livelihoods impossible for the weak supply chain players. In the case of water consumed in food supply chains, farmers facing serious market and environmental volatility manage about 90 % of the 92 % of water consumed in food supply chains. In the circumstances in which they operate they cannot provide the water ecosystem stewardship services that the economies and societies of southern Africa require. These sub-optimal systems characterise the water, energy and food nexus world-wide. An additional reason for analysing the water, energy and food nexus from a supply chains perspective is that those who operate these market and public systems are deeply informed on what is politically feasible and can identify the reforms that are politically feasible as well as needed.

The term nexus has been adopted in this Chapter because it captures the attention of both professionals and scientists as well as non-professionals. It suggests that the complexities of water, energy and food provision are susceptible to scientific and rational analysis. In addition, it implies that these complex processes can be modelled and that the management of the three elements of the nexus—water, energy and food—can be comprehensively optimized. In practice the most appropriate modeling tool—namely Integrated Assessment Modelling (IAMs)—has not been deployed. The contrast with climate change science is stark. In climate change science at least 17 laboratories have developed climate change IAMs. They contradict each other. As a consequence it has become normal to publish results that are the outcomes of putting the metrics from four or more IAMs together to provide an ensemble model that integrates (averages) the separate outcomes. The focus of this climate science is temperature, a very narrow focus indeed, compared with the optimization challenges of modeling the water, energy and food nexus. It is not surprising that the huge budgets devoted to researching temperature have not been mobilized to model the much more complex nexus that provides water and energy services and food. Such a budget is unlikely to be mobilized. It will be argued here, however, that a sub-optimal version of the nexus already exists in market supply chains in the systems of provision in which water energy and food are delivered. More important it will be argued that those operating these supply chains will be more likely to provide relevant metrics than those conceptualising the nexus and its complex interactions.

The Chapter adopts this different, practitioner and practice, point of departure with the term nexus providing the overall frame. Within this frame it is recognized that there exists a complex operational system of private sector markets subject to public policy and regulation. Neither the markets nor the public regulation are ideal. But they operate, are evident and are researchable. They are managed by professionals and public sector officials with immense knowledge, information systems and expertise. Albeit operating in separate and non-integrated market silos and attendant public sector policies and interventions.

These operational systems deliver water services and energy services in rough and ready food supply chains that exist in rough and ready markets which are subject to very highly politicized consumer, private sector and public policy interests. These interests have—especially over the past 60 years—tended to align and privilege the interests of consumers via, for example, cheap food policies and the interests of powerful corporates especially in the international food and energy supply chains.

This Chapter therefore highlights and explores the international relationships—and their national level consequences—between three *operational* sub-nexi—the *water and sewage services supply sub-nexus, the energy sub-nexus and the food sub-nexus.* All three interact and are impacted by two other major global processes. One of them is natural, namely *climate change*, which is a major local and global factor in both the energy and food supply chains. The other is *trade* which plays a major role in the mitigation of water scarcity by enabling access to food and virtual water for about 150 out of the 210 economies world-wide (Kivela 2013).

It is further suggested here that those running the water, energy and food supply chains are already deeply involved in detailed interactive processes, mainly via commercial contracts, that together have come to be recognized as a sub-optimal nexus. Unfortunately, the accounting rules that determine the way these transactions are made only capture some inputs and impacts. They do not capture the costs of water for example and they incorporate misleading costs of energy. Those who manage and mismanage these supply chains know their way about their sub-optimised supply chains, and they know about—and in many cases endure in for example the case of farmers—the consequences of these sub-optimising practices. They are also well equipped to identify the hot-spots that have become evident risks to their operations and livelihoods (SABMiller 2014) but they are not incentivised by public policy or by price signals in their respective market to operate sustainably according to measures of economic or ecosystem security.

There are other reasons for the sub-optimal outcomes of the operational supply chains. Unrecognised scarcity and politics are very important. The ways that natural resources such as water and energy have been developed and consumed by society in southern Africa and internationally have *not* been shaped by awareness of their scarcity or their value. As a consequence the current generation of scientists, legislators, and those managing private sector supply chains have a special responsibility. They find themselves coping with market failures and unsustainable systems. At the same time legislators do not have the political space to install regulations and accounting rules to steer the markets towards sustainable behaviour. Reversing the assumptions that natural resources come free and that there are no consequences from misuse is a challenge that politicians, understandably, handle with very great difficulty. Radical change is not currently politically feasible at the international level as evidenced by the travails of World Trade Organisation negotiation of subsidies and the value of ecosystem protection (Melendez-Ortiz et al. 2009).

The main impact of international processes on the southern African economies is global trade. By the mid-twentieth century major, often very long international private sector supply chains had been established. These long international supply chains, in which water and energy are traded as commodities or as embedded inputs, have existed for over a century. They have supplemented, in major ways, the short private sector sub-national supply chains that have secured food and energy for the world's increasingly urbanised populations. Most of the world's 220 or so national economies are net food importers (Kivela 2013), despite some of these economies having significant land and water resource endowments. This condition is a good example of poverty determining water poverty. Water poverty does not determine poverty.

The SADC economies are mainly net food-water (Virtual Water) importers despite some of them having significant land and water resource endowments (see Fig. 1). South Africa is the biggest participant in SADC international food-commodity trade, accounting for 32 % of SADC exports and 40 % of SADC imports. These numbers strongly influence the regional trade metrics especially those that record the value of the traded commodities. The farms and vineyards of South Africa produce high value crops which are also water-efficient. As a consequence, Virtual Water exports from SADC, while volumetrically less than half the imports, are net earners of foreign exchange.

At this point in history, at the beginning of the twenty-first century, the outcome of the past two hundred year evolution of the political economies of food (Paarlberg 2013) and of energy (Pascual and Elkind 2009) is that global food prices are determined by massive policy interventions by the OECD economies. These OECD policies also determine in some cases, and in others have a deep influence on, the prices in emerging and especially in developing economies. They have impacts on the SADC economies which make livelihoods difficult for SADC farmers from the tendency of international food prices to fall in the 1961–2003 period. The rapid falls in food prices after the volatile price hiatus of 2008–2014 have also been damaging

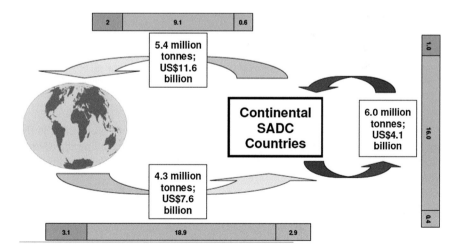

Fig. 1 Imports and exports of agricultural products in 2012 amongst the continental SADC countries and the rest of the world (Food commodity tonnages and values are shown in *text boxes*; the associated virtual water transfers are shown as *bars* with *blue*, *green* and *grey water* in cubic kilometres per year). *Source* Phillips (2014)

to farmers. These distortions have been reinforced by World Trade Organization arrangements as well as by their absence. Water resources have been dangerously impacted by these conditions because it has throughout been assumed that there were no costs associated with over-using, depleting and polluting unvalued or under-valued water ecosystem services and the atmosphere.

It will be a recurring theme in this analysis that the international supply chains that impact the economies of southern Africa scarcity values of the water and energy embedded in food and manufactured commodities, are not reflected in the prices paid by consumers for the goods they purchase in private sector markets. Nor are the use values of water and energy captured. In addition, the exchange values along the supply chains have been very severely distorted by subsidies and taxes. As a consequence, the costs of degrading water resources, the atmosphere and other ecosystem services cannot be signalled in the prices consumers pay for food and energy.

Populations is another factor that must be taken into account in evaluating the current natural resource consumption and managing practices in the SADC economies and the international forces that determine current and future water and energy consumption options. Both the African and global contexts are relevant. For the past decade demographers have been reassessing population projections (UN-DESA 2015; Connor 2015). The latest United Nations Division of Economic and Social Affairs (UN-DESA) 2015 global population projections are challenging especially for Africa.

> With the highest rate of population growth, Africa is expected to account for more than half of the world's population growth between 2015 and 2050. During this period, the populations of 28 African countries are projected to more than double, and by 2100, ten African countries are projected to have increased by at least a factor of five: Angola (SADC), Burundi, Democratic Republic of Congo, Malawi, Mali, Niger, Somalia, Uganda, United Republic of Tanzania and Zambia.[1]
>
> The concentration of population growth in the poorest countries presents its own set of challenges, making it more difficult to eradicate poverty and inequality, to combat hunger and malnutrition, and to expand educational enrolment and health systems, all of which are crucial to the success of the new sustainable development agenda (John Wilmoth, Director of the Population Division in the UN's Department of Economic and Social Affairs UN-DESA 2015).

The estimate that gets most attention in the 2015 UN-DESA projections is the suggestion that the population of the African continent—including the North African countries could reach 4.3 billion by 2100. This total would exceed the 2100 estimate for Asia.

SADC will be affected by these demographics. The SADC economies exist alongside sub-Saharan economies that are currently net food importers. The new population estimates indicate that these sub-Saharan economies will struggle to meet their future much higher food requirements with their own natural endowments. SADC economies cannot look to Africa for their food and closely related water security. They will have to get whatever they can afford on global markets.

---

[1]Note: five out of the ten listed economies are SADC members.

The prospects for water and food security are very concerning. Future energy scenarios are difficult to evaluate because it is not clear to what extent and how soon the potentially rich renewable energy resources of the SADC economies can be mobilised.

2 Conceptual and Operational Issues in the International Domain

This following discussion will highlight the heightened awareness internationally amongst scientists and resources managing professionals of the ways that the over-consumption and poor stewardship of natural resources impact the sustainability of three key strategic supply chains—(1) water and sewage services (2) energy services, and (3) food. These analysts have, for the past five years, been attempting to identify and conceptualise a nexus of optimum and mutually beneficial—or at least less damaging options.

What follows is not a comprehensive review of the literature on the water-food trade nexus, the energy-climate change sub-nexus and what is being called in this discussion the *grand nexus*. Rather, it is a brief attempt to observe, analyse and contribute to their further conceptualisation. In 2008, the World Economic Forum (Waughray 2011), in the activities of which the author began to contribute as a member of its World Economic Forum Water Advisory Council (Allan 2011a), identified water, food, energy and climate change as a nexus. It was concluded (Waughray 2011) that business as usual in consuming water and energy to underpin supply chains could very seriously multiply the risks for society, the economy and the environment world-wide. There was convincing hot-spot evidence of the consequences of the high risk behaviour in OECD, BRICS (Brazil, Russia, India, China, South Africa) and developing economies. These hotspots were present in every continent but most prominently in the United States, Central Asia, across India and northern and western China as well as Australia.

In order to mobilise support and to demonstrate that current unsustainable behaviour could be mitigated by changing the ways that water and energy were allocated and managed, the World Economic Forum Initiative needed a term that captured the attention of those managing and mismanaging many interacting systems. The term adopted had to capture the notion of linkage, of the interacting impacts of water and energy consumption, of mutuality and of shared values in market systems. Unfortunately, there were no accounting or reporting to ensure that all the essential costs were captured that would bring first, effective water and energy saving outcomes and secondly, ecosystem protecting outcomes.

Identifying the gap—in general terms—between business as usual and an efficient and effective mode of resource consumption was relatively easy. But defining systems and sub-systems that would be susceptible to modelling has proved to be beyond the capacity of the research community. More important the private sector corporates that do have the research capacity to develop metrics that could contribute to integrated assessment models have not felt any pressure, or sensed any reputational risk, that would make it necessary to come to terms with the nexus idea and populate it with metrics.

Experience shows that the private sector international corporates tend only to engage with a potentially paradigm changing idea if they see that the approach could generate metrics that might expose them to criticism on the ways they could be misallocating and mismanaging natural resources—such as water and energy. There is a recent example of how they can respond to a new concept. The narrative of Virtual Water and water foot-print metrics reveals how the corporates tend to operate. Initially in the case of Virtual Water and water footprints they observed the emergence of the notion of embedded water but judged they needed to take no action. They reacted very swiftly, however, to the publication of water foot-print metrics generated by the Water Footprint Network in 2001 (Hoekstra and Hung 2002). They realised that they needed to be better informed than their potential critics and quickly funded detailed examinations of their own operations and of the products they processed and manufactured. They generated very comprehensive and also much more detailed metrics on water footprints than the Water Footprint Network. They also deployed the more demanding method of life cycle analysis (LCA).

The outcome has been that the corporates have been able to use their detailed foot-print and life cycle analysis data to reassure their critics that they have significantly improved the way they consume water, and also energy, within their factories and processing plants. They quickly recognised, however, that the volumes of water consumed in their factories represented a tiny proportion of the water consumed by their suppliers on farms world-wide. In a few cases, a corporate worked with an international non-governmental organisation (NGO). The World Wildlife Fund (WWF) is the major international NGO that has contributed to the development metrics on water resource consumption in supply chains. Their work and publications are relevant to southern Africa as a significant part of it was carried out jointly with SABMiller which has a strong presence in southern Africa as well as globally (WWF\SABMiller 2010).

The nexus narrative has moved only a short way compared with the 25 year narrative of the Virtual Water where conceptualisation was followed by the foot-print metrics phase. The three phase process of (1) concept (2) metrics and (3) the recognition by supply chain corporates to respond to the possibility of reputational risk of not coming to terms with the concept has not taken place in the short time that the nexus idea has been in currency.

There may never be any metrics that galvanise Corporate Social Responsibility (CSR) departments to brief their respective Chief Executive Officers (CEOs) about imminent reputational risks. To date the process of developing integrated assessment models has proved to be too conceptually challenging, too costly and too demanding of the limited research capacity available in the science and NGO communities. The corporates that have the capacity to fund such modelling, are unlikely to mobilise the research until they sense risks to their reputations. It is also unlikely that reporting and accounting rules that capture the value of natural resources and the costs of mismanaging them will be installed for some decades.

For the southern African economies this discussion of modelling the nexus and accounting for natural resources is important because it shows that the nexus concept for the moment has only highlighted a problem. The idea has also focused the attention of diverse analysts such as the authors of this volume. But only very approximate nexus metrics have been mobilised and there has been no mobilisation of the research power of the private sector players that dominate the water, energy and food supply chains. These same players also hold much of the relevant data needed to populate such models.

It is concluded that the survival of the nexus explanatory framework depends on the extent to which its advocates can incorporate the rich explanatory evidence that exists in water, energy and food supply chains that are currently natural resource blind.

In another potentially important nexus promoting international initiative in the 2010 and 2012 period three departments of State of the Federal Republic of Germany convened a suite of international nexus meetings attended by numerous scientists, as well as by government, private sector and NGO professionals. The courteous convenors allowed me to contribute a videoed critique of the frenetic engagement on the nexus at that time (Allan 2011b). In this video, I argued that a profound and useful conceptualization of the grand nexus was lacking and also suggested that the absence of an over-arching theoretical frame was making it impossible for those attempting to engage to communicate effectively. I emphasised that the sub-nexus of water-food-trade had been effectively conceptualized. But the energy-climate change sub-nexus and a theorisation of the water-food and the energy sub-nexi into a comprehensive nexus had not been achieved. In addition, I also argued that the operations of private sector food supply chains and energy supply chains had determined priorities as well as the blind-spots. These blind-spots were integral to the business-as-usual sub- and grand nexus operations where water and energy were being managed and mismanaged. In addition, it was also noted that these major supply chain players would be the agents that could most effec-tively analyse, and constructively engage, to address the contradictions that were becoming evident as a consequence of the attempts to develop a grand nexus approach.

It has been shown that water, energy and food supply chains have been shaped in major ways by private sector markets and international trade. These markets are inadequately regulated because they do not have accounting rules that generate market signals that incentivise resource stewardship and sustainable and ethical investment. Water for example is not valued in food supply chains. There are many casualties in these highly distorted systems of provision. For example, farm livelihoods are impoverished by low food prices and farmers cannot afford to provide essential ecosystem stewardship services to regulate the supply chains.

Those who operate the private sector supply chains of *food* and *energy* are deeply acquainted with the challenges in the sub-nexus in which they respectively operate. They are, and will be in future, the agents that know most about their respective sub-nexus operations and constraints. They will also be the agents who could potentially identify and adapt to the risks of ignoring the necessary mutualities and dangerous contradictions of current grand nexus of the political economy. In the international system farmers, big-agricultural corporates and other major corporates in food processing, food marketing, super-markets, oil and gas operations, vehicle manufacture and transportation will all be key players. Subsidies, taxes and the poorly informed choices by consumers in both sub-nexi supply chains will also play significant roles in steering decision-making that involve water and energy. They do this in the as yet separate uncoordinated sub-nexi supply chains identified in Fig. 2.

The *international* grand nexus of water-food-energy—in which the nexus of water, energy and food play a very important role—is populated by players operating market mechanisms and supply value chains. These supply chains in the grand nexus are not yet equipped to expose effectively the environmental and social risks

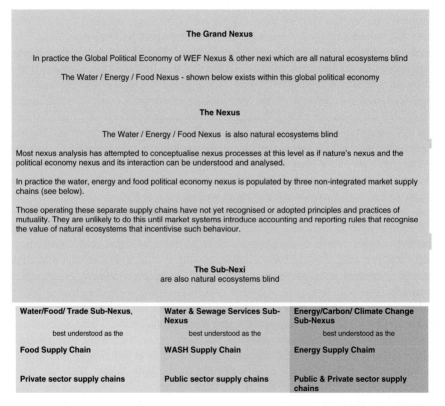

Fig. 2 Supply chain version of the nexus idea (Showing how operational supply chains are foundational. At the same time the Nexus insofar as it can be conceptualised is actually part of global and local political economies that operate supply chains. The water, energy and food supply chains are just three of many operational supply chains). *Source* Allan and Matthews (2016)

associated with business as usual. This impediment to the effective coordination of natural resource use exists because these market systems fatally lack reporting and accounting rules for water. Voluntary measures, with international impacts, are emerging for energy and to a lesser extent for water. Such rules, for example the approach proposed by the Carbon Disclosure Project (CDP 2013), signal to those who manage vulnerable and sometimes very scarce water and energy resources the consequences of not grasping the significance of natural resource scarcities and values. Market players, especially at the international level have risked the planet, profits and society as well as the sustainability of their own enterprises because they have assumed that the atmosphere and water ecosystems will forever take the strain. They are proving to be particularly blind to the consequences of the inadequacy of the accounting rules with which they comply.

It is understandable that it has proved to be easier to recognise the threats to, and to the crises over, water and energy availability for their own local operations. These processing and manufacturing plants need relatively small, but secure, volumes of blue water. The southern economies of SADC along with scores of economies in semi-arid and arid regions world-wide have provided evidence that corporates have recognised their vulnerabilities to local water scarcity (Hausmann 2010a, b; Miles and Jones 1994; Nestlé 2007, 2010, 2014; SABMiller 2014; WBCSD 2013). They are not yet willing to recognise fully their collective role in accelerating global water resource security and the sustainability of natural ecosystems. They recognise and respond to operational risks to their own operations but they are not incentivised or regulated by existing market accounting systems to make allocative and management decisions that would mitigate their collective impacts on the ecosystems on which they depend. There are no accounting rules. These corporations are often managed at the highest level by accountants. But they align with the short-termism of the market, which is prone to doing the wrong things spectacularly well (Drucker 1967).

A number of very important fundamentals are obscured, and therefore ignored, as a consequence of existing levels of knowledge on the parts of government, corporates and consumers. For example Fig. 3 illustrates a number of very important knowable unknowns. It shows that separate food and energy market systems consume very different proportions of embedded water and embedded energy to deliver food and energy to consumers.

Figure 3 identifies the food supply chain associated with the provision of food and fibre and the non-food supply chains involving energy. Those managing these supply chains have not systematically engaged to optimise the consumption of water and energy. For example, they have not engaged, over the risks in food supply chains—both economic and environmental—of consuming vast volumes of not properly costed water. The embedded energy associated with pumping water for crop and livestock production is not captured in current accounting systems. Nor have they engaged over the real, but unaccounted, costs of water and energy associated with the *wasteful* allocation and management of water and energy in the production, conveyance and consumption of goods and services. Most significantly the benefits of adopting a wider understanding of competition and of mutuality in

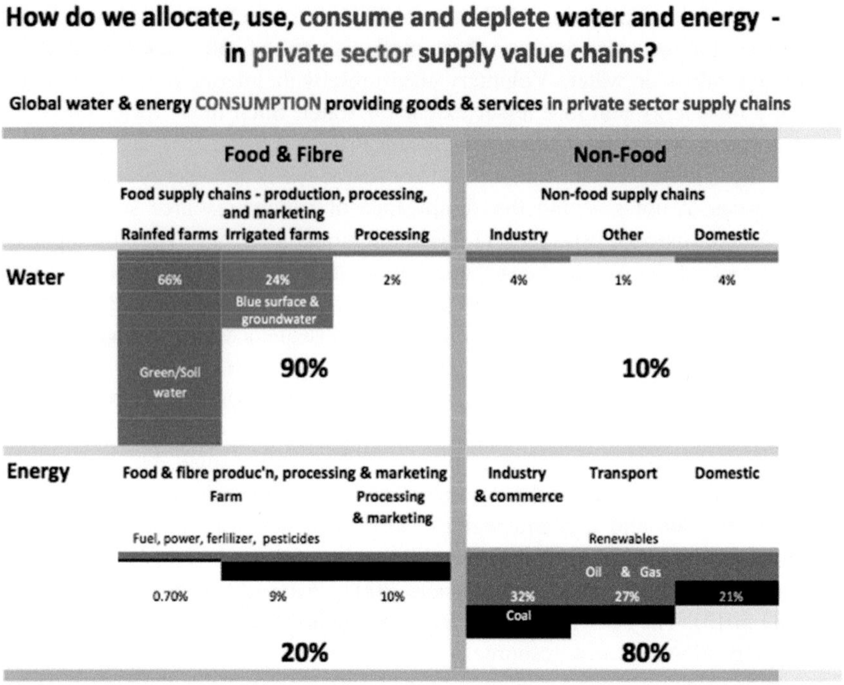

How do we allocate, use, consume and deplete water and energy - in private sector supply value chains?

Global water & energy CONSUMPTION providing goods & services in private sector supply chains

	Food & Fibre			Non-Food		
	Food supply chains - production, processing, and marketing			Non-food supply chains		
	Rainfed farms	Irrigated farms	Processing	Industry	Other	Domestic
Water	66%	24% Blue surface & groundwater	2%	4%	1%	4%
	Green/Soil water	**90%**		**10%**		
Energy	Food & fibre produc'n, processing & marketing			Industry & commerce	Transport	Domestic
		Farm	Processing & marketing		Renewables	
	Fuel, power, ferlilizer, pesticides				Oil & Gas	
	0.70%	9%	10%	32% Coal	27%	21%
	20%			**80%**		

Fig. 3 Global asymmetries. *Sources* Mekonnen and Hoekstra (2011), British Petroleum (2013)

resource use with respect to profit have not been operationalised. Profit is foregrounded. People and the planet are not (Elkington 1994). Reporting and accounting systems focus on profit and not on human and ecosystem health.

The SADC economies are subordinate to these international profit motivated market supply chains. Profit, as currently defined in capitalist market economies can never be a sound metric to incentivise effective stewardship of natural ecosystems. Profit is, unfortunately, the main market incentive for the market players who deliver water intensive food and energy intensive services. It is currently the duty of corporate CEOs to generate profits so that shareholders—the owners of their companies—get the best possible return. Responsible approaches to people and the planet are also needed but those who operate the supply chains can only protect natural ecosystems to the extent that existing inadequate regulatory regimes require them to do so.

The discourse is changing at the international level but high profile leadership from private sector CEOs of major corporates is occasional and not universal. The interventions by the potentially very important accounting industry are painfully hesitant. Accounting practitioners have devoted a lot of energy to understanding the ways in which accounting practices are inadequate. But they have not yet devoted much resource to establishing effective accounting systems that would properly regulate the consumption of water and energy according to triple bottom line principles. As yet there are very few reporting and disclosure rules and no accounting rules for water by which to steer. High level international bodies have been established such as the Sustainability Accounting Standards Board (SASB 2015) and the International Federation of Accountants (IFAC 2015) is active and projects essential messages. IFAC has recently highlighted the need for accountants to step up and account for water (Allan et al. 2015).

In the discussion so far our inadequacy in fixing threats to water, energy and food security has been established. One of the main reasons that the necessary market and regulatory reforms have not been installed is because there is a poor understanding of the level of risk being taken, first, on the part of politicians who have the means to regulate markets, secondly, on the part of key senior market leaders, and thirdly, most importantly, on the part of consumers. They all need to be aware of the proportions of water and energy consumed in different sectors of the world's economies revealed in Fig. 3. Figure 3 also reveals some major asymmetries in the use of water by economic sector in the two sub-nexi of water-food-trade and of energy-climate change.

The lack of awareness is universal. Consumers and those who operate in food supply chains and even water scientists and professionals are unaware that about 90 % of water consumed by society is devoted to food production and provision. Less than 10 % of water is consumed by society in its industries and for domestic services, and much of that water is in practice recycled. There are some important international examples of what is possible in the recycling and manufacture of blue water. For example, in Singapore and Israel, almost all domestic and industrial water is recycled. Urban water could potentially be recycled in most economies. All the water used for the production of non-food commodities and services is blue water. Blue water is the water we can divert, pump, convey and relatively easily value. As pointed out in Chapter "Mechanisms to Influence Water Allocations on a Regional or National Basis" it is the water that is has to some extent been licensed and allocated.

Figure 3 also highlights the significance of the consumption of green water in the production of food and fibre. Green water is the water held in the root zone, sometimes called effective rainfall. Its consumption is not priced. In areas where green water can be supplemented by blue water irrigation is possible.

The most important revelation of Fig. 3 is the extraordinary allocative asymmetries of water consumption compared with energy consumption in the food and non-food sectors. The approximate 90/10 ratio in water consumption for food sector/non-food contrasts with the 20/80 ratio for the consumption of energy in the food sector/non-food sector.

Blue water is available in rivers, lakes, reservoirs and in underground aquifers. Blue water can be diverted, pumped, conveyed and polluted and is the main concern of hydraulic engineers and of water resources and water quality scientists. Blue water can also be valued and even priced. It is usually priced in non-food uses but very frequently mispriced. Blue water, however, is very rarely properly priced as food-water.

Green water comprises most of the water used to produce food and fibre in private sector food supply chains. Of the 90 % of water in food supply chains about 70 % is green water, although this varies from place to place and from product to product (Chapter "Quantifying Virtual Water Flows in the 12 Continental Countries of SADC"). Green water is the water held in the soil profile after rainfall. Natural vegetation and crops both draw on this water which they transpire to the atmosphere. All this food-water consumption is lost to local users. It is recycled to the atmosphere in the global hydrological system. It cannot be recycled locally. Green water is a major and essential resource in food production and it also provides important ecosystem services.

Blue water consumed in agriculture is different hydrologically and economically from green water. Blue water when used for irrigation competes with other blue water consumers in industry, in other services like electricity production, in the provision of recreational amenity, in the provision of domestic water services and in environmental flows. The very vulnerable ecosystem services of blue water are increasingly highly prized by society but their value is not yet captured. Poor farmers cannot afford to steward them. Other farmers are not incentivised in market system to protect them.

Providing irrigation infrastructure and related funding to improve farm livelihoods is only sound to the extent that the blue water resource can be sustainably consumed over the long term. Sustainable blue water consumption for irrigation is usually underpinned by water allocation and licensing processes. In SADC countries as is the case world-wide, these processes are inadequate (see Chapter "Mechanisms to Influence Water Allocations on a Regional or National Basis").

In the SADC region and internationally the key role of farmers as the managers and stewards of huge volumes of water is not respectfully recognized by engineers, water scientists, economists, or by society and governments more generally. Food consumers are especially under-informed and unappreciative on the role of farmers in managing and stewarding water and energy. International market systems that connect consumers with blue and green water are dysfunctional in terms of informing consumers on the impacts of their food choices on water resources and water ecosystems.

Food supply chains are very complex systems. In many ways they are very effective systems. They deliver vast volumes of food in long international and short local supply chains. However they are rough and ready in the sense that they recognize many, but not all, of the inputs in very selective and very partial accounting systems. Misleading contracted prices connect the agents in our food supply chain markets.

Figure 3 has highlighted the very significant asymmetries in the use of water and energy in two of the major global private sector supply chains—food and energy. There are other asymmetries that are possibly even more important in impacting any attempt to integrate or synergise the two sub-nexi that are the focus of this analysis. Figure 4 is a preliminary attempt to conceptualise the different types of water and energy available. It highlights first, the nature of the renewable and non-renewable water and energy resources in the respective supply chains and secondly, it

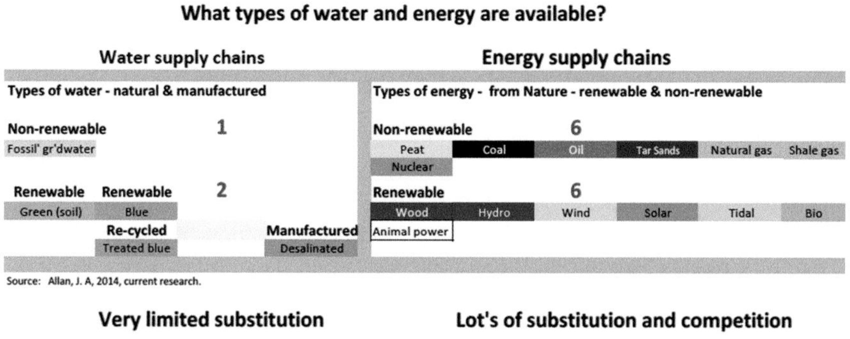

Fig. 4 Difference in supply chains (Water supply chains are very different from energy supply chains. There are only *two types of renewable natural water—green* and *blue*. There is one type of *non-renewable blue water—namely fossil groundwater*, which is a very minor source of water overall amounting to less than 0.01 % of water consumption world-wide. Although it is very important in some desert economies such as Libya and Saudi Arabia. Volumes of manufactured (desalinated) water are also as yet almost invisibly small. In contrast, society has mobilized more than a dozen types of energy. There are at least *six types of renewable energy and at least six types of non-renewable energy*. The two supply chains are especially different in relation to *substitutability*. In water supply chains the substitution for water is unusual. In energy supply chains the substitution for energy is normal)

illustrates the stark differences with respect to diversity and substitutability of water sources—for which they are few, and of energy—for which there are many.

The main message of Fig. 4 is that there are very different resource managing options and very different competitive contexts in the water and in the energy markets both globally and at the SADC scale. The energy sector is characterised by very competitive markets as a consequence of rapidly changing technologies that affect the costs of production of the numerous non-renewable and renewable sources of energy. Water, on the other hand, can only be green renewable, blue renewable with small volumes of non-renewable blue water (sometimes called fossil water) and even smaller volumes of manufactured water. Substitutability is not a simple or a safe option in the water sector. Blue water can be used to supplement green water but everywhere such measures have proved to be problematic or extremely damaging for water ecosystems.

3   The Consequences of the Highly Politicised Provision of Water Services, Energy Services and Especially of Food Provision

Because water, energy and food are vital and therefore very easily politicised at the national level all three supply chains have been impacted and distorted in major ways by government subsidies and related payments. In these highly politicised contexts there are some very powerful players both nationally, but especially internationally.

J.A. Allan

They are powerful because they are very deeply informed, they have significant contractual and commercial leverage and they operate at a scale that can influence public policy in the economies in which they operate, such as the SADC economies, as well be responsive to public policy. The asymmetric power relations make sustainable practices impossible to practice by the weak supply chain players such as farmers. In food supply chains farmers face extreme market and environmental volatilities. In these destabilising circumstances they manage about 90 % of the 92 % of water consumed by an economy in food consumption (Hoekstra 2012).

SADC farmers cannot provide the water ecosystem stewardship services that the economies and societies of southern Africa require. These sub-optimal systems are also evident in the water, energy and food nexus world-wide. It has been noted above that an additional reason for analysing the water, energy and food nexus from a supply chains perspectives is that those who operate these market and public systems are deeply informed on what is politically feasible and can identify the reforms that are possible as well as needed.

Figure 5 shows a widely recognized way of understanding the structure of a neo-liberal society and provides an easy to communicate approach to analysing advanced political economies. Other political economies currently referred to as emerging and developing, to a substantial extent can also be understood via the lens of the four ways of life approach. Figure 5 identifies three—what social theorists (unhelpfully) call—*social solidarities* that do things to, and for, a fourth solidarity, namely *civil society*. The three solidarities that do things to civil society are—first the *public* state, secondly, the *private* markets and thirdly, collective *civil movements*—such as unions, NGOs and other activists. The latter do not have the hard power and administrative powers that have been acquired by the state. Or market power. They have the soft power deriving from advocacy and collective action.

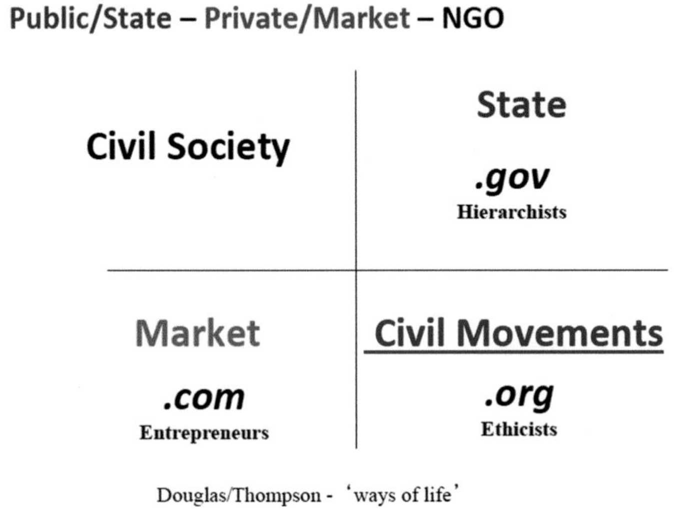

Public/State – Private/Market – NGO

| Civil Society | State
.gov
Hierarchists |
| Market
.com
Entrepreneurs | Civil Movements
.org
Ethicists |

Douglas/Thompson - 'ways of life'

Fig. 5 The three social solidarities. *Source* Thompson (1990)

Civil society is all of us at breakfast time after which we go out to work in public and private sector organisations. Others contribute to civil movements and to union and religious activities which are often inspired by rights and environmental activist causes. Human rights, rights to water, rights to livelihoods and the right of the environment to protection are prominent concerns in this solidarity.

Such civil movement advocacy explains a disproportionate amount of the progress being made in ecosystem protection, especially internationally. Private sector bodies that deliver food and energy could take the lead in protecting the atmosphere and water ecosystems but they are only required to comply with reporting and accounting systems that privilege the delivery of under-priced food and energy. Accounting rules ensure that the costs of most inputs are captured. However, the contribution of natural ecosystems is not accounted. In competitive markets there is no incentive currently to highlight the contradiction of the business as usual approach that seriously damages natural ecosystems. NGOs, and especially international NGOs, deploy advocacy to get private sector operators to do what non-existent reporting and accounting rules cannot achieve.

The structure set out in Fig. 5 is useful as one of the sub-nexi, or supply chains, considered in this study can be mapped on to it. Food supply chains consume over 90 % of water need by an economy. The diagram provides the underlying frame for

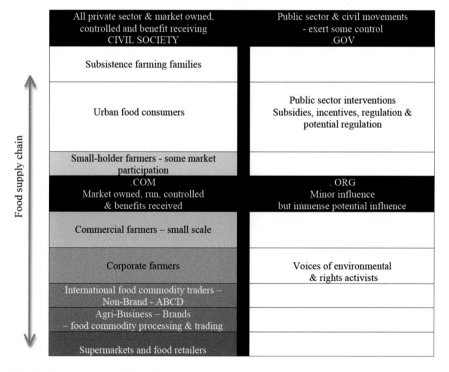

Fig. 6 The four ways of life with the private sector food supply sub-nexus mapped on to it

Fig. 6 which shows how the water-food-trade private sector market supply chain—or sub-nexus—can been mapped very effectively on to the four ways of life structure shown in Fig. 5. Private sector food supply chains occupy the left hand side of the diagram.

The food supply chains are easily related to by all the players who engage in them and/or benefit from them even if they are blind to the many dangerous accounting problems that reflect their preference for cheap food ahead of secure ecosystems. These conditions are global as well as local. So any region that is a significant food commodity importer such as SADC cannot rely in the long term on a world system which is designed to deliver cheap food at all costs.

One of the asymmetries identified above—the massive volumes of water devoted to food and fibre production can also be illustrated. Figure 6 framed on the four ways of life concept shows how the 90 % of the water consumed by society via private sector food supply chains can be mapped on the four ways of life space. It is the *water-food-trade sub-nexus* that is being mapped on to the civil society and private sector space. *Farmers* manage, mismanage and consume 90 % of this water in crop production. This supply chain operates as a market but in the absence of the necessary reporting and accounting rules it is water value and externalities blind.

Figure 7 also shows that the small volumes of non-food water provided in water and sewage services are handled mainly by public sector service providers, and often through formal allocation processes (Chapter "Mechanisms to Influence Water Allocations on a Regional or National Basis"). Local authorities and other municipal bodies provide domestic and industrial water services everywhere except in the UK. Private sector companies, many of which are major global corporates

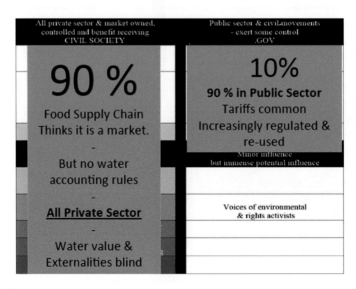

Fig. 7 Allocation of water resources

only provide between 10 and 20 % of domestic water and sewage services together with industrial water provision world-wide. The non-food water and sewage services sector has a chequered history internationally which is not the concern of this analysis. Privatisation of these services is gaining ground but only gradually. There is no sign that there will be other than a gradual increase in the level of privatised management of water and sewage services.

The global water, energy and food supply chains are distinguished in one further way. Their histories are different and these histories have affected all the world's economies including those of what have become the SADC economies. The global food provisioning system (Friedmann 1993; McMichael 2009) and the global energy provisioning systems, which it is necessary to review here have differed especially in the past half century.

The global energy regime was dominated until 1972 by the alliance of three governments—the United States with the UK and Netherlands, and with seven oil companies which were known as the Seven Sisters. Five were US corporations. BP and Shell were UK and Netherlands based. The oil price shocks of the 1970s changed the global power relations as the national oil companies of the producer economies were able to increase substantially their proportional take of the financial returns. The OECD economies should have read the signals four decades ago and have taken the opportunity to develop renewable energy. There would have been a twofold benefit. First, renewable energy technologies would by now be affordable. Secondly, we would have avoided four decades of full on degradation of the atmosphere.

Energy prices have fallen rapidly following the 2008 and 2011 price spikes as they did in the 1980s after the 1979 spike. Food commodity prices—that track global energy prices—are also continuing to fall. For staple food commodities such as wheat they were half their 2010 levels by 2014. These volatile international processes have been beyond the control of the SADC economies and this situation will continue. The consequences of falling food prices will negatively affect SADC farmers and commodity price volatility will continue to be very difficult to mitigate in the SDADC economies generally. In the energy sector there is an urgent need to promote renewable solar and wind energy.

The global food regime has also been dominated for even longer by the United States (McMichael 2009) than was the global energy regime. The 1970s price spikes affected both food as well as energy. The global food system has been dominated by another alliance of the US State and its private sector trading corporates since the second half of the 19th century. The ABCD grain and livestock trading global corporates (ADM (US), Bunge (US), Cargill (US) and Dreyfus (French)) have been transacting about 70 % of global food trade by volume for the past five decades. They have been joined by other commodity traders such as Glencore based in Switzerland in the past two decades. In the last decade, however, there has been a major change of international significance with the emergence of East and South-East international traders.

These recent major shifts in the global food commodity trading regime have in some ways been equivalent in their impact on international food trading as the global

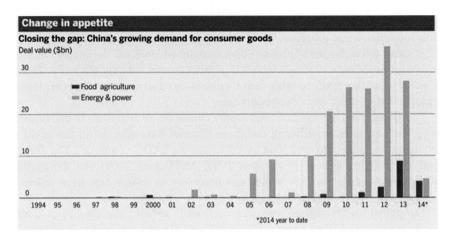

Fig. 8 China—global corporate acquisitions in energy and food/agriculture—1994–2014. *Source* Bloomberg (2014)

energy price shocks of the 1970s did on the power relations in the global oil regime. The changes in the international power relations of the global food regime have been brought about by the activities of East Asian corporate traders. Keulertz (2013) identified a new group of global traders in East and South-East Asia—the NOW corporates. Noble, the 'N' in the NOW, was a Hong Kong based trader with revenues as big as ADM by 2013. Noble has recently been acquired by Chinese corporate state investors—COFCO. Such acquisitions are part of a trend and reflect a shift in the global political economy. The trends shown in Fig. 8 confirm the phenomenon and reflect the strategy of China with respect to food and energy (Bloomberg 2014). These international supply chains and the leverage which they can exert are a major focus of this analysis. There are significant upward trends in the international acquisition of corporate food and energy interests by China. Meanwhile China is not investing significantly in water and land for food production in Africa (Brautigam 2012).

Hassan and Thiam (2015) provide a very useful analysis of recent trends in the international trade of South Africa—SADC's major international trader. Their analysis exposes two trends which are driven by international pressures and international opportunities. These trends are likely to continue.

These two very significant international trends in the 2002–2013 period are first, the increase in the exports of higher value horticultural commodities shown in the Virtual Water 'exports' evident in Table 1. Secondly, Table 2 shows that there have been very significant and consistent trends in the relationship of South Africa with global markets. Table 2 shows that South Africa's exports are steadily rising to Asia and the rest of the world. There is a reciprocal decline in exports to America and the EU. In the same period the proportion of imports from America and the EU have fallen at a higher rate than exports. Those from Asia and the rest of the world have also increased at the higher rate. South African trade with the rest of SADC does not demonstrate these trends.

Table 1 Trends in South Africa's virtual water 'trade' in billion m^3—2002–2013

Year	Total	AOR1C	HORTIC	CROPS	OTHRAGRIC	MANUF	Other
2002	1.17	1.06	1.76	−1.05	0.35	−0.84	0.95
2003	1.31	1.31	1.88	−0.81	0.24	−0.97	0.97
2004	1.00	1.12	2.08	−1.11	0.15	−1.43	1.31
2005	1.21	1.54	2.17	−0.68	0.05	−2.05	1.72
2006	−0.02	1.05	2.15	−1.10	0.00	−2.89	1.82
2007	−0.08	1.02	2.82	−1.82	0.01	−3.47	2.38
2008	1.27	2.41	3.57	−1.28	0.11	−4.18	3.04
2009	2.32	2.68	3.70	−1.15	0.13	−2.86	2.50
2010	3.18	3.04	4.25	−1.30	0.10	−2.86	2.99
2011	2.77	3.00	4.46	−1.36	−0.11	−3.84	3.61
2012	0.62	2.29	5.08	−2.58	−0.22	−5.37	3.70
2013	2.06	4.58	6.76	−2.08	−0.10	−6.60	4.08

Key: *AGRIC* Agriculture, *HORTIC* Horticulture, *OTHRAGRIC* Other Agricultural, *MANUFAC* Manufacturing
Source Hassan and Thiam (2015). Used (Quantec 2014)

Table 2 Destination and origin of trade flows with South Africa (shares of exports and imports 2002–2013)

Year	Exports				Imports			
	Africa				Africa			
	Total (%)	SADC share (%)	America and EU (%)	Asia and ROW (%)	Total (%)	SADC share (%)	America and EU (%)	Asia and ROW (%)
2002	14.96	71.50	44.70	40.34	4.13	49.15	58.43	37.44
2003	14.52	68.95	43.01	42.47	3.69	58.57	57.48	38.83
2004	12.75	67.11	44.51	42.74	4.59	53.73	53.88	41.54
2005	13.77	67.40	44.85	41.37	5.16	66.92	50.73	44.11
2006	13.04	66.45	45.20	41.77	6.91	49.34	47.38	45.71
2007	13.54	65.31	44.62	41.85	7.89	67.63	47.05	45.07
2008	15.49	68.60	42.26	42.25	8.54	68.11	44.74	46.73
2009	16.93	66.93	34.43	48.64	8.00	55.85	45.08	46.92
2010	14.96	70.00	35.36	49.68	7.83	59.43	44.01	48.15
2011	14.91	71.14	33.67	51.42	7.68	55.22	43.50	48.82
2012	17.58	73.63	32.53	49. 89	9.89	56.66	40.67	49.45
2013	18.38	74.54	31.87	49.74	9.38	47.37	39.85	50.77

Source Hassan and Thiam (2015). Calculated using Quantec (2014) Easy International Trade database

Those managing the SADC economies must be savvy concerning these global trends and shifts in power. They must understand the interests of the powerful international food corporates in the food supply chains and of the powerful international energy corporates as well as those of the national oil companies in the energy supply chains. They must recognise that food markets and energy markets are separate and not integrated. They are separate markets or supply chains; operating as separate sub-nexi in the language of the nexus. If oil companies owned irrigated farms and if farmers were intimately involved in mining there would be an automatic awareness of unsustainable competitive utilisation of water and of its different values in alternative uses. Their accountants would be aware of and alert to profit impacting contradictions such as asset depletion and asset impairment.

It should come as no surprise that private sector investors play a very important role in keeping in place the business as usual distortions and impediments to change. Private sector *investors* invest in the separate supply chains reinforcing the ways that the supply chains of water, energy and food are managed. Some, so called ethical investors have noted the contradictions of the separateness of the supply chains. Responsible investors are emerging internationally and in SADC. This ethical investment is small as yet but potentially influential, both as a consequence of their investing power as well as because they are beginning to coordinate their responsible and environmentally considerate investment behaviour (Yach et al. 2001; Greczy et al. 2005).

In the absence of a grand market nexus tracked by accountants the ad hoc and poorly informed hot-spots approach will prevail in the places where resource availability crises make engagement unavoidable over water, food and energy. When the hotspots become evident—usually too late to remedy easily—outcomes will always be very highly politicised. Hotspots have, for example, become evident as a consequence of the contradictions with respect to impacts on water resource associated with developing shale gas and tar/oil sands as well as in producing ethanol from corn and other food crops. This hot-spots approach will continue until the private sector players, who operate supply chains recognise that there are unacceptable risks in not promoting policies that take into account the essential mutuality of water security and energy security.

4 Global Commodity Prices and What Importing Economies Can Do in Very Uncertain Circumstances About Volatility and Long Term Trends

It is necessary to devote a section of this chapter to global commodity prices. Serious problems have come about as a consequence of long term trends internationally that have made ever cheaper food with occasional periods of major price volatility. These trends in international food prices have tracked quite closely international energy prices. It has been shown above that existing political and commercial incentives have driven farmers, traders, food manufacturers and supermarkets to deliver cheap food. Consumers have become addicted to it. Politicians do not know how to reverse the trend.

Advances in technical efficiency, especially on farms, have played a role in reducing food prices as have not accounting for the costs of environmental inputs

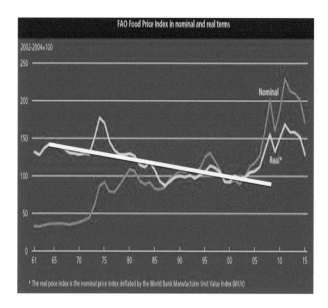

Fig. 9 Trends in the global price for food (1960 to 2010—showing the tendency for prices to fall except in periods of global instability in energy and finance. The major price spike in the 1970s was closely related to oil prices. The price rise beginning in 2003 was also oil price related. The major spikes of 2008 and 2011 were brought about by the dual impacts of oil price rises and by the financial crises of 2008 and 2011. Food prices are falling again but it is too early to say where they will settle when the world economy, oil prices, inflation and interest rates return to what has become regarded as normal). *Source* FAO and OECD, 2015

and of impaired ecosystem services. Figure 9 shows how successful the drive to provide cheap food has been since 1960. As noted food prices have throughout this period very closely tracked energy prices. Unfortunately the way food prices track energy prices is—as yet very poorly understood. The link cannot be assumed to be evidence of the nexus between the two supply chains.

Figure 9 illustrates the scale of impact of the oil price hiatus of the 1970s that caused food prices to spike in that decade. Success in making food cheaper has for the past half century made it impossible to foreground the idea that the environment also needed to be stewarded and accounted for. The expectation of cheap food has made consumers blind to the unsustainable tendencies of private sector food supply chains.

Food prices are still falling at the time of writing—2015. Pessimists—thinking about the sustainability of ecosystems—want the current higher prices to shift the political discourse to foreground stewardship. But the path dependence determined by subsidies in the OECD countries will tend to overwhelm this brief window of potential opportunity for policy change. The targeted public payments in the US Farm Bill and in the Common Agricultural Policy which very powerfully impacted food prices are vey difficult to change. The scale of the 2014 US Farm Bill will, over the next decade, put very close to one trillion dollars into food supply chains. $780 billion will be for food stamps (Washington Post 2014). The annual food stamp budget in the US is $78 billion annually. Only 60 countries in the world have total annual GDPs in excess of this figure. The coming decade has international policies locked in place that will continue to make it difficult for consumers to detect the underlying environmental and economic fundamentals.

Will society pay the farmers to steward ecosystems as well as produce enough food? There are two main ways to answer this question. The most politically difficult path would be to include the costs of stewardship and environmental protection in the price of food. A second way, less politically difficult, would be to introduce new payments for stewardship. The EU Common Agricultural Policy and the US Farm Bill have both initiated such payments but at a scale and a pace determined by complex legislative and farm level politics. At present stewardship is primarily underpinned by Corporate Social Responsibility (see Chapter "Virtual Water and the Private Sector").

5   Concluding Comments

One of the roles of international assistance should be seen as a remedy to the damage done to the farm sectors in developing economies by the OECD dominated food trade regime.

International conditions and contexts are very important for the SADC region as it is economically dependent on the global system for both energy and food for its economic security. Energy and food security are very emotional issues which are highly politicised. In this chapter a number of international conditions that will determine SADC's options in consuming and stewarding water and energy resources have been highlighted.

First, food is very emotional and food prices are extremely highly politicised. These fraught politics affect international food supply chains as the provision of cheap staple food is an international imperative. In addition, the livelihoods of rural voters in developing countries, and in BRICS economies, are significant. Secondly, there are serious asymmetric power relations in food supply chains. Farmers are generally weak everywhere. But farm lobbies in OECD countries are very powerful and the resulting OECD agricultural policies—such as the EU CAP (Common Agricultural Policy) and the US Farm Bill—have led to low global staple food prices. This was the case during the five decades before the recent volatility that started in 2008. OECD food production systems produce cheap staple food the price of which does not reflect the true costs of production. This condition makes it impossible for would-be commercial farmers in regions where they operate with primitive technology and basic agronomic practices to have viable livelihoods. Governments always authorise the import of readily available under-priced commodities from the world market. Generation after generation of subsistence farmers have been forced back into poverty in regions such as SADC. Commercial farmers have also been negatively impacted by this cycle that exposes them to unfair global competition. Thirdly, those with power in private sector food supply chains, especially international corporates, usually handle a very small proportion of the embedded water resources that are consumed inside the fence of their operations. They also have potential contractual leverage over farmers who do manage the vast volumes of water in the supply chains in which they transact with farmers. As yet the international corporates have had few incentives or no prompts from accounting rules to engage outside their warehouses, silos, factories and wineries. This position is, however, is showing signs of changing as Pegram points out in Chapter "Virtual Water and the Private Sector". Fourthly, market signals and the reporting and accounting systems that track them are dangerously partial, as well as blind to the values of natural resource inputs, especially of water. Any progress towards a sound approach to managing natural resources requires the constructive engagement of private sector corporates with their suppliers and with governance processes more generally (Muller 2013).

Water and energy are essential natural resources for the SADC economies. Secure access to regional and international water and energy is an existential issue. International markets and international political conditions are major factors in determining SADC economic security. Water and energy sustain societies, economies and ecosystems. But not currently in ways that are necessarily sustainable. This analysis has shown first, that the SADC region is economically dependent on international food and energy systems and is subject to some difficult long term trends in market conditions. It also has to cope with the international market and its price volatilities that characterise key supply chains. These market systems providing commodities and services are separately organised and they do not yet, nor will they easily, align with what could be a rational nexus approach to their allocation and management.

One of the purposes of the chapters has been to evaluate the role of the nexus approach to allocating and managing increasingly scare water and energy in the SADC economies. It has been concluded that the approach is very difficult to deploy comprehensively and there is no evidence on the international stage that the approach is robust either conceptually or operationally. It has been argued here that there is no evidence that there has been, or will be, a significant effort to fund the necessary integrated assessment modelling that would reveal its complexity. More importantly, it has been suggested that a pragmatic approach to coordinating water, energy and food, capitalising on the operational information systems that exist in long established supply chains is being installed. Those operating in these supply chains are expert in how the global political economies of the water, energy and food provision operate. They understand the impacts of subsidies, taxes, mispricing as well as the frailties of the economic and of the environmental regulatory regimes. Finally, they know how the potential efficiencies from foregrounding and implementing measures that consume water and energy in ways that reduce the overall consumption of both could be implemented—even if they are unwilling as yet to adopt them. Most important they know why they are not being implemented.

In other words the problem of adopting nexus type measures is not the availability of expertise. Rather it is the strong path dependent behaviour that characterises local and especially international water, energy and food supply chains. The path dependence is especially strong with respect to expectations on the part of consumers/voters on how commodities and services are priced. Cheap food and cheap energy are very highly politicised issues.

The international supply chains are dangerously dysfunctional in two other areas. The first relates to the expectations of private sector investors. Their expectations and preferences reinforce the separateness of the local and global food and non-food supply chains. Food supply chains are wholly in the private sector. Energy supply chains are substantially in the private sector. These two supply chains are subject to the expectations and pressures inimical to sustainable practices exerted by the short term investor priorities. They want high returns quickly. The current culture of the investor community does not have the long term perspective necessary to steward scarce natural resources such as water and energy.

The other major impact from outside the region is the consequence of the inadequate public sector regulatory practices in the dominant OECD economies. In addition to the price distortions established by OECD governments they have also kept in place weak environmental stewardship regimes that in turn influence international practice. The past two decades have witnessed some positive shifts in legislation and some talk of adopting shared values on the part of big corporates. But the impact of such shifts is as yet inadequate.

The global demand for energy and especially food are very concerning. Recent global statistics on population (UN-DES 2015) suggest that the current pace of population growth in Africa is an international factor that will impact the SADC economies. Ten African countries are projected to increase their populations by at least a factor of five by 2100: Angola, Burundi, Democratic Republic of Congo, Malawi, Mali, Niger, Somalia, Uganda, United Republic of Tanzania and Zambia. Five out of the ten listed economies listed are SADC members. The SADC economies are likely to remain net food importers for the rest of the century.

The international perspective taken in this analysis has highlighted how the global regimes that provide food and energy operate in ways that make it especially difficult for farmers to be responsible stewards of nature's ecosystems. This condition is a universal global problem. The poor farmers of the SADC economies are very seriously affected by these global food and energy systems and have no power to change them. The dominant global food and energy regimes have evolved from the alliance of US and EU Governments with the Big-Ag, Big-Ag-Trade and Big-Energy corporates. These global systems are natural resource value blind. These long established regimes are being challenged by investors and traders in East Asia. The SADC economies should watch these Pacific Rim initiatives very carefully.

To enable communities and farmers to acquire infrastructure that cannot otherwise be afforded is sound. But it is a remedy to a problem caused by the UK Government and the other OECD Governments in that they play such an important role in determining the long term trends in food prices. The US Farm Bill and the EU CAP (Common Agricultural Policy) which have played such big parts in causing the declining trend in global food prices are for SADC a problem that need to be addressed. Poor farmers in less capable developing economies cannot invest to improve their livelihoods in their impoverished circumstances as long as this global cheap food regime continues to be kept in place OECD and US policies. Solutions actually lie in the corridors of DC,

Brussels and Whitehall where decisions that would impact the ecologically dysfunctional food and energy supply chains could be reformed. Recognition of the need to assist emerging and developing economies with substantial investment in the protection of the atmosphere is evident in the announcement by the UK Government that it would devote almost $90 billion on tackling climate change in poor countries over the five years from 2015 (Guardian 2015).

The experience of the SADC region is similar to that of other regions of Africa. The problem of the yield gaps cannot be effectively addressed because of the problems of mobilizing investment in farming. This investment needs to be local and national and delivered by responsible international investors. Crop and livestock yields could be trebled and more in the better watered northern SADC economies (Morris et al. 2012; University of Nebraska 2015), but would require investment in transport and other infrastructure (Chapters "The Future of SADC: An Investigation into the Non-political Drivers of Change and Regional Integration" and "Virtual Water and the Nexus in National Development Planning"). Responsible inward investment in agriculture has been much talked about recently but despite the rhetoric the record to date has been slow and mixed with respect to outcomes (Allan et al. 2012). China's strategies are appropriate. Their investment is in infrastructure, usually in major structures such as ports, roads, power and communications. Such infrastructure enables future responsible and effective farm investment in green and blue water that will make land productive. No doubt they will do that as it becomes evident that commercially viable food supply chains can be installed.

In the opinion of this author both the scale and the focus of the Chinese approach will have the most sustainable impact. It will establish the necessary conditions for the expansion of environmentally and economically resilient agricultural rural livelihoods. It will enhance the impacts of all farm and rural investment including that by CRIDF. The author is aware that not all of Chinese investment has been successful and the SADC region has experienced problems in the energy sector (see Chapter "Electrical Power Planning in SADC and the role of the Southern African Power Pool").

Finally, any nexus approach that aspires to have significant impact requires effective reporting and accounting rules to guide investors and shape on-farm behaviour. This study has emphasized the consequences of the absence of effective reporting and accounting on ecosystem use. For those who use the term nexus want it to have an impact it would help if they could make the term more than an arm-waving moment in conference presentations.

References

Allan T (2011a) The water, energy and food security, Interview at the Federal Government of Germany conference in Bonn—16–18 November 2011. http://www.water-energy-food.org/en/news/view__225/tony-allan-the-tension-between-sustainability-and-intensification-captures-totally-what-this-meeting-is-about.html. Accessed 4 July 2014

Allan JA (2011b) Soft approaches to sustainable intensification for water security. In: Wauhray D (ed) Water security: the water, food, energy crisis, The World Economic Forum Water Initiative

Allan JA, Matthews N (2016) The water, energy and food nexus and ecosystems: the political economy of food and non-food supply chains. In: Dodds F, Bertram J (eds) The water, food, energy and climate nexus: challenges and an agenda for action. Routledge, London

Allan JA, Keulertz M, Sojamo S, Warner J (2012) Handbook of land and water grabs in Africa: foreign direct investment food and water security. Routledge, London

Allan T, Keulertz M, Colman T (2015) The complexity and urgency of water: time for the accountancy profession to step up. International Federation of Accountants (IFAC), New York. http://www.ifac.org/global-knowledge-gateway/viewpoints/complexity-and-urgency-water-time-accountancy-profession-step. Accessed 20 Sep 2015

Bloomberg (2014) Food replacing oil as China M&A target of choice. http://www.bloomberg.com/news/2014-05-30/food-replacing-oil-as-china-m-a-commodity-of-choice-commodities.html 4 July 2014

Brautigam D (2012) Chinese engagement in African agriculture. In: Allan JA, Keulertz M, Sojamo S, Warner J (eds) Handbook of Land and water grabs in Africa: foreign direct investment food and water security. Routledge, London

British Petroleum (2013) Water in the energy industry: an introduction, London. http://www.bp.com/content/dam/bp/pdf/sustainability/group-reports/BP-ESC-water-handbook.pdf

CDP (2013) Use of internal carbon price by companies as incentive and strategic planning tool. CDP North America, New York

Connor S (2015) The fertile continent: Africa's population to grow explosively, The world in 2100. The Independent, London

Drucker P (1967) The effective executive: the definitive guide to getting the right things done. Harper Collins, New York (2002 edition)

Elkington J (1994) Enter the triple bottom line. In: Henriques A, Richardson J (eds) The triple bottom line; does it all add up? Earthscan, London

FAO & OECD (2015) Food price index in nominal and real terms. FAO, Rome

Friedmann H (1993) The political economy of food: a global crisis. New Left Rev 197:29–57

Greczy GC, Stambaugh RF, Levin D (2005) Investing in socially responsible mutual funds. Wharton School, Pennsylvania

Guardian (2015) UK will spend £6bn of foreign aid budget on climate change, The Guardian, 28 Sept 2015, p 8

Hassan R, Thiam DR (2015) Implications of water policy reforms for virtual water trade between South Africa and its trade partners: Economy-wide approach. Water Policy 17(4):649–663

Hausmann C (2010a) Opening keynote—supply crisis or green revolution? Address presented at Agriculture Outlook Europe 2010, London

Hausmann C (2010b) Bringing externalities inside: incorporating sustainability into our business model. Address presented at National Conference for Agribusiness in Purdue University, West Lafayette

Hoekstra AY, Hung PQ (2002) Virtual water: a quantification of virtual water flows between nations in relation to international crop trade, value of water research Report Series, No 11. Water Footprint Network, Delft

Keulertz M (2013) Drivers and impacts of farmland investment in Sudan: water and the range of choice in Jordan and Qatar, Unpublished Ph.D. thesis at King's College London

Kivela M (2013) Virtual water 'flows' in international food trade: mapping the net virtual water 'imports' and 'exports' of 210 economies, Unpublished Master's Dissertation, King's College London

McMichael P (2009) A food regime genealogy. J Peasant Stud 36

Mekonnen MM, Hoekstra AY (2011) National water footprint accounts, value of water research Report Number 50. Enschede: Water Footprint Network

Melendez-Ortiz R, Bellmannn C, Hepburn J (2009) Agricultural subsidies in the WTO green box: ensuring coherence with sustainable development goals. Cambridge University Press, Cambridge

Miles L, Jones M (1994) The prospects for corporate governance operating as a vehicle for social change in South Africa. Heinonline, Johannesburg

Morris M, Binswanger HP, Byerlee D (2012) Awakening Africa's sleeping giant: prospects for commercial agriculture in the Guinea savannah zone and beyond, World Bank and FAO

Muller M (2013) Water security: beyond prices and markets to values, governance and government, World Economic Forum—Global Advisory Council on Water https://www.academia.edu/5840132/Water_Security_Beyond_prices_and_markets_to_values_governance_and_governments. Accessed 4 July 2014

Nestlé (2007) The Nestle water management report (Rep.). Nestle, Geneva

Nestlé (2010) The Nestlé policy on environmental sustainability. Nestec Ltd, Vevey

Nestlé (2014) Nestlé S.A website (online). Available at: http://www.nestle.com/. Accessed 10 June 2014

Paarlberg R (2013) Food politics. Oxford University Press, New York and Oxford

Pascual C, Elkind J (2009) Energy security: economics, politics strategies and implications. Brookings Institution, Washington DC

Phillips D (2014) Quantifying virtual water flows in the 12 continental countries of SADC, position paper, opportunities to support the uptake of virtual water concepts: Project 1807, Version 2 June 2014. CRIDF, Pretoria

Quantec (2014) Easy International Trade database, Pretoria. Available from http://www.quantec.co.za

SABMiller (2014) Make beer use less water. http://www.sabmiller.com/files/reports/positionpaper_water.pdf 5 July 2014

SASB (2015) The sustainability accounting standards board. See http://www.sasb.org/. Accessed 20 Sept 2015

Thompson M (1990) Cultural theory. Westview Press, Boulder

UN-DESA (2015) The world population prospects: 2015 revision. Division of Economic and Social Affairs, New York 29 July 2015

University of Nebraska (2015) Global yield gap atlas. University of Nebraska and Wagenigen University, Lincoln Nebraska. http://www.yieldgap.org/. Accessed 29 Sept 2015

Washington Post (2014) $950 billion Farm Bill in one diagram, Washington Post, Washington DC 28 Jan 2014. http://www.washingtonpost.com/blogs/wonkblog/wp/2014/01/28/the-950-billion-farm-bill-inonechart/?utm_source=feedburner&utm_medium=feed&utm_campaign=Feed%3A+Counterparties+%28Counterparties%29. Accessed 4 July 2014

Waughray D (2011) Water security: the water, food, energy crisis. The World Economic Forum Water Initiative

World Business Council for Sustainable Development (2013) Water, food and energy nexus challenges. WBCSD, Geneva. http://wwwf.org.uk/wwf_articles.cfm?unewsid=4204. Accessed 16 Sept 2015

WWF\SABMiller (2010) Tackling water scarcity together. WWF and SABMiller, Woking

Yach D, Brinchmann S, Bellet S (2001) Healthy investments. J Bus Ethics 3:191–198. http://link.springer.com/article/10.1023/A:1017518819776#page-2. Accessed 5 July 2014

Embedding the Virtual Water Concept in SADC

Reginald M. Tekateka

Abstract This chapter explores how the concept of Virtual Water can be embedded in SADC and its institutions. An argument will be made that the unpacking of Virtual Water may impose challenges on public servants in communicating and that the concept must position itself as a satisfactory alternative to the infrastructure development goal that most of SADC aspires towards. Importantly, also, countries that are relatively well-endowed with water and land would have to devote resources to producing goods for trade inside SADC, rather than to the rest of the world. An astute multi-tier communications strategy aimed at broad dissemination of the benefits, while recognising the legitimate sovereign concerns will therefore be required. It will be helpful to target countries individually as well as collectively at the basin and regional levels within SADC. Another important element of the strategy will be to put Virtual Water squarely on the SADC agenda. This can be done through bodies and forums such as the Global Water Partnership's Multi-stakeholder Forum, the Water Resources Technical Committee of the SADC Water Sector, the SADC River Basin Organisations' (RBO) Forum and targeted individual RBO's, the Water Strategy Reference Group (WSRG) and the SADC Water Ministers.

1 Introduction

In the late 90s up to and including its high profile treatment at the World Water Forum in Kyoto, Japan in 2003, Virtual Water was largely seen as a highly academic proposition that offered little by way of contribution to the key identified challenges of the water sector in the region and the achievement of regional economic integration. Virtual Water was largely seen as failing to convince in its

R.M. Tekateka (✉)
Steering Committee of the Global Water Partnership, Maseru, Lesotho
e-mail: tekatekar2@gmail.com

© Springer International Publishing Switzerland 2016   201
A. Entholzner and C. Reeve (eds.), *Building Climate Resilience*
through Virtual Water and Nexus Thinking in the Southern African
Development Community, Springer Water, DOI 10.1007/978-3-319-28464-4_9

ability to help bridge the infrastructure gap that led to economic water stress that is typical of much of the SADC region. Soon after its conception leading into the new millennium, Virtual Water (VW) became the after-hours pub debate topic among senior water sector officials attending regional meetings of the SADC water sector. To some it was seen as a fascinating academic issue which, however, had little relevance to the solution of the problems of the day. To others it was an important topic that required deeper understanding in order that its relevance is appreciated. The words of a Tanzanian colleague during one of these discussions late into the night still ring audibly. "Never come to my village and tell my people about water they can never drink". This assessment prevailed at a time when the region was particularly receptive to progressive ideas that promoted cooperation around water resources management and sharing.

This forward looking regionalist stance that was led by a well-knit inner group of water ministers of the time was not able to eradicate strong national sovereign impulses that continued to influence country choices often in conflict with agreed collective approaches. This remains the case in river basin and regional spheres alike. Also there remained a strong sense of unease at the relatively high level of infrastructure development enjoyed by some countries. Any proposal that appeared to favour the more highly developed rather than level the playing field was not likely to elicit enthusiastic support. Virtual Water, like any new concept in the water sector is likely to be assessed against its ability to address these concerns while enhancing water's role in regional development.

It should be noted that the concept emerged at a time when the main focus of the water sector was on how it should be strengthened such that it could meaningfully contribute to regional integration and poverty alleviation in southern Africa. To many, VW did not appear to reflect this. Moreover there was a lot on people's minds. The region had barely completed the process of formulating and agreeing on a regional water policy and strategy, on internalizing and agreeing on the applicability of the key principles of the United Nations Convention on the Law of the Non-Navigational Uses of International Watercourses, formulating and agreeing on the SADC Water Protocol, negotiating and ratifying river basin agreements, participating in the formulation of a regional contribution to the World Water Vision and going through the difficult process of rolling out and implementing the revised regional structures of a recently transformed SADC.

Within this context, the initial introduction of the VW concept found very little traction in SADC.

2 Background

The prospect of water becoming a sector in its own right free from the shadows of agriculture and the environment created much excitement in the region and fuelled the fire of regionalism within the sector. Movements such as the Southern African Regional Technical Advisory Committee (SARTAC) thrived in this new dispensation. A small but activist group of water ministers emerged in the region that was personally and collectively committed to the regional approach, literally staking their reputations in support. They appreciated the limitations of going it alone as countries, seeing the benefits of basin approaches and regionalism. They also understood how cooperation around water resources management and development could support regional integration. They therefore welcomed the fact that SADC had created the unprecedented opportunity for water to play this all important role. This inner group of ministers was to play an important role in galvanizing the energies of officials, SADC staff and even civil society in support of this regional project.

In turn this led to a similar development among middle level and senior government officials in the sector. Officials felt empowered to work towards the success of the vision and in several instances bureaucratic corners were cut with the assurance that there would be backing from a significant number of ministers. It became quite difficult to oppose a regional idea in SADC water meetings. Over time this group was to break up due to normal attrition and it was not in all cases that replacement ministers or officials had similar zeal.

Nonetheless the lesson learnt from this period was the important role that the "group within the group" can play in leading and energizing cooperation around an issue of regional importance. The challenge today is to consider how such a group of champions can be forged to promote acceptance of VW across relevant sectors.

3 Infrastructure Development as a Regional Priority

The need to develop water infrastructure to address the imbalances of access and equity as well as energy development were already among the key issues preoccupying regional governments during this period. Infrastructure development was to become the mantra of the African Ministerial Council on Water (AMCOW), established in 2002, led to a large extent by a group of SADC Water Ministers who had had a head start thanks to the processes cited above. The key test for any initiative at the time was whether it supported water infrastructure development. The establishment of the African Water Facility (AWF) later to be housed at the African Development Bank (AfDB) and aimed at catalyzing infrastructure development was the direct result of this burning desire on the part of AMCOW Member States. The decision to house the AWF at the AfDB was taken in the hope that this would create the necessary confidence on the part of potential donors to contribute to the facility.

However, even after this decision on its location was made, many still felt that it would become encumbered by the bureaucratic regulatory framework that is typical of international financial institutions, and thus slow down access to the urgently required funding. This was despite the fact that Africa was never able to propose a credible alternative location for the AWF. Many African observers saw as a sinister betrayal the "sudden decision" by the European Union to establish its own separate

European Water Facility (EUWF) in parallel while purporting to work towards the same goal as AMCOW. Policy responses to the recommendations of the World Commission on Dams were also formulated on the basis of the extent to which the recommendations were deemed not to be in conflict with this goal. Many African Ministers openly described the Commission's work as a ploy to slow down Africa's development or divert Africa away from its chosen development path.

Other regional dynamics also played a part, fuelled in part by attitudes attributable to the southern Africa region's still very recent apartheid past. The relatively high levels of development by South Africa of two key rivers in the region; the Orange-Senqu and the Limpopo rivers (two Pivotal basins as outlined in Chapter "The Future of SADC: An Investigation into the Non-political Drivers of Change and Regional Integration"), whose benefits were seen by some as accruing only to South Africa, and doubts about South Africa's willingness to be a team player in the region were important aspects of this. Some in the water sector welcomed the high profile that water enjoyed in the new South Africa and that country's strong leadership in the processes leading up to the formation of the SADC water sector. Others were suspicious of South Africa's motives even to the extent of feeling that Pretoria intended to take over the stewardship of the nascent sector at the expense of Lesotho. A lot of time and resources have gone into addressing this unease about South Africa's motives as manifested in several basin wide projects, some ongoing, aimed at confidence building. While South Africa was lauded for the lead role it was playing in promoting regional cooperation in the sector including through initiating and hosting ministerial meetings to this end, her motives remained suspect. Zimbabwe was also seen by some as having had a head start that needed to be taken into account as the region moved forward together.

An observation that begs to be made in this regard is that the question of suspicion around the motives of the new South Africa with particular reference to the water sector has still not been raised openly 20 years later. What this ultimately indicates is that no serious collective attempt to address it, will likely be made in the foreseeable future, and it remains a sensitive political matter which should not be tackled directly. Nor as a result will South Africa itself fully appreciate the need or how best to work harder in reassuring her sister states of her good neighbourly intentions.

The other unfortunate reality is that too often a regional initiative can become subject to suspicion as soon as it is seen to benefit South Africa in any way. In fairness it is also understandable that countries that are lagging behind in the development journey would feel that regional initiatives should be aimed at primarily assisting them in catching up with those historically more developed. It should not be forgotten that countries forego sovereign rights in order to promote sovereign interests. This quest for regional equity is a political reality that will not likely just go away. Importantly, however, this must be seen in the context that South Africa, as demonstrated in most of the other chapters, is a pivotal state in economic, electricity, trade and water matters.

4 Sovereign Considerations

While purist pundits of shared management of and equitable access to trans-boundary water resources have continued to stress the need for riparian states to accept sacrifices in regard to national sovereignty, basin politics have continued to remind us that we are still some way away from that 'perfect state' even within a fast paced SADC water sector. Zambia stood firm throughout the years often to the frustration of fellow Zambezi Basin partners as well as sector officials keen to see an important upstream country ratify the basin agreement of one of the region's potentially most transformative rivers. It remains to be seen whether the decision to designate a senior Zambian water official as the Executive Secretary of the Interim Secretariat of the Zambezi Watercourse Commission (ZAMCOM)—now Permanent Secretariat—contributed in any way towards the change in position on the part of Zambia that has led to ratification. Or indeed whether the appointment of another Zambian in that position in the permanent Secretariat will promote the resolution of what is seen as one of the first tasks of ZAMCOM, the determination of approaches towards the reasonable and equitable use of the waters of the basin.

Moves to develop the transportation potential of the Shire River a tributary of the Zambezi, has been frustratingly slow, and this has spilt over into agreements to establish electricity transmission lines between Malawi and Mozambique (see Chapter "Electrical Power Planning in SADC and the Role of the Southern African Power Pool"). While it would seem unimaginable that either of these infrastructure development options will not eventually happen, for now sovereignty and environmental considerations between Malawi and Mozambique have held back progress in this regard.

Elsewhere, Botswana with its rather robust stance against Namibia's wish to draw more water from the Okavango to meet essential water supply needs, also adopts a national interest stance, albeit similarly citing environmental concerns.

The demise of the SADC Tribunal raises concern, not so much because a Member State felt exposed and betrayed by the system,[1] but rather because recovery has proved so difficult to achieve. In fact the decision to disband the Tribunal was, as per the Rules of Procedure of the Summit, unanimous. How this is reflective of the regional appetite for integration remains to be seen, although it is not necessarily a foregone conclusion that the region is politically moving away from integration principles. For now bilateral solutions are the only recourse on offer to resolve disputes in the water sector contrary to regional agreements.

There is certainly still no evidence that countries will deliberately sacrifice the national interest for the regional goal of integration, particularly in the light of the need for rapid development. From the evidence of the relative progress achieved in the establishment of river basin organisations (RBOs) it is evident that they will

---

[1]Indeed, the specification in the SADC Treaty that Tribunal decisions should be 'final and binding' (Article 16.5), would seem to be unusual in international treaty terms.

embrace the collaboration option where national gain can be demonstrated, although in many cases the countries are opting to limit the powers and functions of these RBOs to an advisory role.

A weakness that one notes with SADC is the absence of a regional ethos among the region's citizens. By design or omission, there has not been a deliberate, and in any way sustained, effort to court the peoples of the region. Where such efforts have been made, it has consisted of isolated and disconnected competitions such as essay writing and sporting events that in their execution do not necessarily promote any sense of belonging. One suspects that this may well remain the case until the people of the region are more intimately involved in its processes and programs. A regional sense of empathy needs to be cultivated such that this can exert influence on governments. Ideally, issues of regional interest could then feature on party platforms in national elections. For the time being countries are in no way compelled to be sensitive to public opinion beyond their respective borders.

The highly visible nature of electricity load shedding, its existing media profile and the benefits of the SAPP and regionalisation (discussed in Chapter "Electrical Power Planning in SADC and the Role of the Southern African Power Pool") may make this a useful platform for this kind of profiling.

5   Virtual Water Concept

The reintroduction of the Virtual Water concept to the region as a possible region-wide strategy may initially be met with scepticism following its initial failure to take root. As regards the water sector, one can also expect that it will be scrutinized against the infrastructure development standard. A critical point here is that the relative complexity of the VW debate may be its undoing. VW does not easily ring true to the average ear. It does not readily demonstrate its benefit as well as its relationship to the infrastructure deficit and debate. As suggested earlier, the majority of the countries of the region consider their situation with regard to water storage and water conveyance infrastructure as woefully inadequate. How does VW address this challenge? Does VW independently bring benefits that would otherwise accrue from increased infrastructure, without bringing with it additional political and risk related costs? Similarly, water abundant countries would also need the assurance that VW would positively contribute to their economic growth opportunities, and that diverting water and land use to trade is indeed the way to go.

The treatment of the subject in the southern Africa region has tended to be in the rather high level intellectual space. This has of course been necessary in attempting to understand VW's essence, implications, its relevance and applicability, as well as how best to adapt it to different political, social and economic scenarios. However we need to look critically into whether a strategy that focuses on the VW concept per se, as opposed to one that uses the concept to demonstrate added value, is likely to achieve optimum buy in. We might also find that we are shifting the burden of defending, unravelling and articulating it onto already overextended government officials. Rather, we may want to consider reverting to already familiar terminology and imagery such as, for instance, "promoting the optimal use of water, land and infrastructure (new and existing) resources, to the benefit of Member States both individually and collectively" (SADC Regional Water Policy) VW will need to be disseminated systematically, broadly and intelligibly if it is to be embraced. To do so effectively will require a multi-tier communication strategy also targeted at specific sectors and interests outside water, in particular but not limited to agriculture, trade, energy, etc. Intelligibility refers to its accessibility to those who will need to understand the dynamics of the farming, trading and other opportunities it would seek to create. Targeting Ministries of Planning (often also held jointly with the finance portfolio) should also be considered. These Ministries in particular have become quite adept at interpreting complex ideas and theories into intelligible and implementable policy options. Nonetheless, it is recommended that the focus of the dialogue should be on using the concept to deliver added benefit, both in terms of providing stronger arguments for financing infrastructure, as well as in economic and environmental benefits to both individual countries and the region as a whole.

There is also merit in considering broad dissemination independently of inter-sector penetration at the very outset. Private print media houses within the region tend to be open to carrying regionally relevant stories and think pieces in their opinion columns. Similarly, Inter-Press Services have shown a willingness to carry transboundary water and regional integration success stories, and could be explored as an option. These wider initiatives will help to broaden understanding and create social demand to nudge political processes forward.

While the initiative is to be rolled out within the context of SADC's core mandate of regional integration it is prudent that the strategy should seek to stress the VW benefits to countries individually. Thus a country by country cost/benefit assessment will need to be developed probably in collaboration with the countries concerned. This has the benefit of carrying the countries along from the onset and in so doing creating demand. In this exercise it would again be a good idea to find a way to involve entities outside government including chambers of commerce, agricultural unions, etc. It will also be wise to shed light on how the region as a whole will benefit. In other words, demonstrate how VW would ultimately strengthen the integration process as well as the integrating economic region itself?

6 The Role of the SADC Secretariat

As a regional body established and mandated to steward regional integration, the SADC Secretariat is going through arguably its most challenging phase since its establishment. It is ill-resourced financially and is consequently grossly understaffed. Moreover, due to its rather limited mandate it has historically tended not to champion independent initiatives or innovations not already endorsed by Member States. However it should be mentioned that whatever its current condition, the Secretariat's core mandate of regional integration remains intact. It is not to SADC

as the agent for change that one turns, but rather as the region's official and duly
mandated regional vehicle for progressive change.

Thus, in order to effectively introduce new thinking or promote new initiatives
including policy options within the region one should tactfully identify entry points
and use SADC's own institutional bodies at various levels or partner institutions to
solicit and cultivate support from Member States and promote the initiative, the
SAPP (Chapter "Electrical Power Planning in SADC and the Role of the Southern
African Power Pool") is a case in point. Often it will be necessary to do spade work
within countries as a parallel process, targeting public and non-governmental
entities where the latter exist, in order to mobilize the necessary support. The aim
here is not necessarily to win support from SADC per se but to have access to and
to utilize its mandated purpose which is to commend duly processed ideas, initiatives and policies for collective approval or adoption by governments. Sometimes
the value of this approach lies in merely assuring governments at country level that
a particular idea, initiative or policy proposal has been adequately considered and
debated and is deemed valid, acceptable and implementable by Member States.
Some recommended entry points include the following:

- Water Resources Technical Committee (WRTC)

The Water Resources Technical Committee is the sector entry point and clearing
house for issues due for eventual consideration by SADC water ministers. Issues are
added to the agenda of this annual meeting through the initiative of Member States,
the Secretariat itself, or through the Secretariat by a recognized sector partner.
Ideally, one would want to have a Member State suggest inclusion on the agenda
with prepared support from one or two Member States. It always helps to have two
or three delegates who are conversant with and supportive and of the issue under
discussion.

- SADC River Basin Organisations Forum

This bi-annual forum also offers a good opportunity to introduce a topic into the
SADC water sector agenda. Its limitation lies in the fact that participants rarely
include people from outside the regional RBO fraternity. It is nonetheless an
important regional forum occupying an increasingly important and influential space
in the region as RBO's grow in cohesion, experience and stature.

- Targeted River Basin Organisations

SADC RBO secretariats have grown in confidence and effectiveness over the
years in their advisory role to governments. They present the opportunity to target a
particular cluster of countries through a trusted medium that is also intimately
familiar with the countries individually and collectively. One carefully identified
and properly cultivated RBO potentially sets up entry into the WRTC. ZAMCOM
may prove to be a useful port of initial call in this regard.

- Global Water Partnership-Southern Africa (GWP-SA)

Collaboration with the Global Water Partnership-Southern Africa (GWP-SA) as a neutral platform could present useful options to reach a broad section of the target audience for the initiative. GWP-SA is a strategic partner of the SADC Water Division. A prime example of this would be use of the often televised GWP Multi-Stakeholder Forum to discuss and debate the strategy. The forum's big advantage lies in its being open to government, non-governmental entities, private sector and any formations interested in participating thus giving it high penetration potential, arguably the broadest on offer across all sectors. This could be co-hosted with GWP with targeted invitations to carefully chosen and sourced panellists including one or two that are not from the traditional regional economic power-houses. Turton has in his Chapter "The Future of SADC: An Investigation into the Non-political Drivers of Change and Regional Integration", postulated a northward shift in hegemony driven by rapidly growing economies and to some extent perhaps less climate vulnerability. The courting of Angola, Mozambique and Zambia in this regard may prove useful. This meeting could be used as the launching pad for a broadening of the regional discussion on virtual water within SADC.

- Water Strategy Reference Group (WSRG)

The WSRG is the forum of all international cooperating partners collaborating with the SADC Water Division. The WSRG's concurrence with and support for the adoption by SADC of a VW based strategy would substantially enhance its credibility. It is conceivable that participating international cooperating partners (ICPs) would also assist in promoting the approach in the other relevant sectors that many of them are also involved in. It would be advisable that care is taken to check if any of the ICPs that wield influence in other sectors are in any way opposed or unconvinced so that this reticence can be addressed early prior to any damage being inflicted. The Department for International Development (DfID) as one of the core funders of transboundary waters in SADC already has a seat at this table, which could be exploited by CRIDF.

- SADC Secretariat

The Secretariat offers an entry opportunity in and of itself as a vehicle for further elaborating the concept and ensuring understanding. Entry into the secretariat also avails the opportunity to use contacts within the Water Division to reach SADC secretariat officials in other relevant divisions and through them access to their respective divisions. The Water Division of the SADC Secretariat plays a crucial role in preparing support documentation for matters put before ministers at their decisive annual meetings. The Division is also responsible for the wording of reports of official meetings. Their willingness to allow collaboration in the preparation of the documentation as well as agreeing to invite independent presenters to official meetings can play a key role in determining the fate of an initiative. Success or failure at this level determines whether or not a matter is approved and recommended for presentation and consideration by the political structures of SADC including the Summit. A matter with relevance for more than one sector such as is the case with VW could stand an even greater chance of success if chaperoned properly.

- SADC Water Ministers' Meeting

The different routes suggested above namely the WRTC which is the technical clearing house for issues eventually presented to ministers, the RBO's with their in-house access to respective ministers, and the SADC Secretariat through its official mandate all give support to and can thus provide a route to the Ministers' agenda. Again it will always help to have cultivated relations with some targeted ministers beforehand while obviously welcoming any others that might independently have found merit in the particular issue at hand.

- Southern African Power Pool (SAPP)

The clear water-, carbon- and cost-benefits to be had by electricity trading through the SAPP makes this SADC institution an ideal point of entry. In this regard, the regular meetings of the various SAPP organs could be targeted for presentation of the core concepts.

7   Conclusion

In order for an arguably radical concept such as Virtual Water to win support within the SADC region, it:

- Needs to be seen as being in line with and in support of, and not in conflict with regionally adopted strategies and principles, chiefly;
- Must support infrastructure delivery, not be seen to be in lieu of it;
- Must promote hydro-supportive growth trajectories;
- Should, however, acknowledge and accommodate the inherent bias towards sovereign interests, while supporting the political goal of integration;
- Must be supportive of the regional desire and priority to develop water infrastructure;
- Must bring demonstrable and quantifiable benefits to countries both individually and collectively;
- Must not exclusively benefit the larger economies but also support the emerging and rapidly growing economies in their aspiration to catch up, recognising the benefits of stronger neighbours and trading partners rather than the risks of competition for resources and markets.

Critically,

- It must avoid appearing too elitist and esoteric in its orientation, specifically targeting the benefits of the concept rather than the concept itself;

- While tactically selective it must be broadly targeted and ultimately inclusive in its roll out;
- The role of the private sector as distinct from but together with the SADC citizenry must be recognized;
- A multi-tier communications strategy will be required;
- It must have a multi-sector orientation; and
- It must show familiarity with key strategies in other relevant sectors.

There needs also be recognition that it will take a while for the concept to grow roots. Mechanisms to sustain the processes through the agencies outlined in the previous section will have to be developed.

Postscript

Charles Reeve

1 The Key Water Related Challenges of the SADC Region

As noted in the SADC Water Policy (SADC, 2006) and elsewhere one of the greatest water challenges facing the SADC region is the variability in water availability over space and time. The drier southern regions have historically responded through the development of large scale hydraulic infrastructure, storing water and transferring it to where it is needed, while the wetter northern regions have not yet fully exploited their relative water abundance through water infrastructure. These northern basins and countries are vulnerable to intra-annual variability and shorter term drought.

2 The Impact of Climate Change

As referenced in a number of papers in this volume climate change is expected to increase this north/south water availability split even further, with the wetter north getting wetter (and perhaps more variable) and the drier south, drier. However it must be noted that the regions food and energy security is dominated by production in the drier South.

3 Drought Stalks SADC Again

Since the majority of the papers in this volume were prepared in early 2014 the SADC region has once again been hit by drought. Analysts' are beginning to speculate that SADC may be facing a drought as bad as or worse than, the worst

C. Reeve (✉)
Climate Resilient Infrastructure Development Facility, Hatfield, Pretoria, South Africa
e-mail: charles.reeve@cridf.com

© Springer International Publishing Switzerland 2016
A. Entholzner and C. Reeve (eds.), *Building Climate Resilience through Virtual Water and Nexus Thinking in the Southern African Development Community*, Springer Water, DOI 10.1007/978-3-319-28464-4_10

drought since records began: 1992. Even before the onset of the summer rains in 2015, much of SADC is desperately short of water. Countries from Tanzania to South Africa have declared drought, and hydropower production is severely affected across the region. Botswana's capital city Gaborone is perilously close to running out of water—the city is on severe rationing and even with the rationing there was a daily deficit of 17 million litres per day on 2nd November 2015 (Water Utilities Corporation, 2015). Tanzania has closed down many of its hydropower plants (BBC, 2015). Zambia and Zimbabwe have severely curtailed power production from the Zambezi River (NBC News, 2015). The World Food Programme is already feeding significant numbers of people in Zimbabwe, and the Government is starting to reach out to its neighbours for assistance (Times Live, 2015). South Africa is bracing itself for emergency action, rainfall is well below average already and temperatures are at record highs (Nomura Economic Insight, 2015). The current water shortage, projected to worsen, is the most pressing challenge for the region. It has already begun to dominate all discourse in public and private communication. It is likely to remain this way in the medium term.

4 Global Economic Challenges Impact SADC

At the time the papers in this volume were prepared three of the SADC countries, Mozambique, Tanzania and Zambia, were amongst the world's ten fastest growing economies in terms of annual average percentage GDP growth. In the period 2001–2010 both Angola and Mozambique were classified in this group. The situation is very different now because of changes in the global economy. Angola, Botswana, Namibia, South Africa, Tanzania, Zambia and Zimbabwe all have mining and mineral export dependent economies; these countries account for 95 % of the SADC GDP. The global fall in commodity prices has very significantly impacted on the economies of these countries. In addition a number of SADC countries have a dependency on sugar exports. This is especially true of Malawi and Swaziland but, other countries such as Mozambique, Tanzania, South Africa and Zambia have a significant dependence. The fall in global sugar prices is impacting the economies of these countries. These global economic challenges mean that the economies of the SADC region are no longer buoyant and this impacts on the plans for infrastructure development together with the regional integration plans of the SADC Regional Economic Community.

5 The Virtual Water Conjecture

It is against this background that the Virtual Water and Nexus perspectives may offer a different view for consideration by national and regional planners around large infrastructure. Alternatives to large scale regional water *transfer* infrastructure, making better use of the *total water footprint* and *Virtual Water trades* in food and electricity across the region may promote regional integration and net benefits for the whole region, without threatening sovereign security, while also promoting growth.

CRIDF is pleased to place the information/evidence contained in this book in the public domain to stimulate and empower discussion through a better understanding of climate resilience issues which will result in better decision-making both for local needs and priorities and global/regional considerations.

References

BBC (2015) Tanzania closing hydropower plants available on http://www.bbc.com/news/world-africa-34491984. Accessed 21 Feb 2016

NBC News (2015) Drought-caused blackouts batter Zambia, Zimbabwe economies available on http://www.nbcnews.com/business/energy/drought-caused-blackouts-batter-zambia-zimbabwe-economies-n454581. Accessed 18 Feb 2016

Nomura Economic Insights (2015) Emerging markets, 'South Africa: water is the new Eskom', 3 Nov 2015

SADC (2006) SADC water policy

Times Live (2015) Zimbabwe government battling to feed its people: vice president Mnangagwa. Available on http://www.timeslive.co.za/africa/2015/10/19/Zimbabwe-government-battling-to-feed-its-people–Vice-President-Mnangagwa. Accessed 18 Feb 2016

Water Utilities Corporation (2015) See, http://www.wuc.bw/tempx/02November.pdf, 2 Nov 2015